Make: mBot for Makers

CONCEIVE, CONSTRUCT, AND CODE YOUR OWN ROBOTS AT HOME OR IN THE CLASSROOM

Rick Schertle
Andrew Carle

Maker Media, Inc.
San Francisco

Printed in Canada.

Published by
Maker Media, Inc.
1700 Montgomery Street, Suite 240
San Francisco, CA 94111

Maker Media books may be purchased for educational, business, or sales promotional use. Online editions are also available for most titles (*safaribooksonline.com*). For more information, contact our corporate/institutional sales department: 800-998-9938 or *corporate@oreilly.com*.

Publisher: Roger Stewart
Editor: Patrick DiJusto
Copy Editor: Elizabeth Campbell, Happenstance Type-O-Rama
Proofreader: Elizabeth Welch, Happenstance Type-O-Rama
Interior Designer and Compositor: Maureen Forys, Happenstance Type-O-Rama
Cover Designer: Maureen Forys, Happenstance Type-O-Rama
Indexer: Valerie Perry, Happenstance Type-O-Rama

December 2017: First Edition

Revision History for the First Edition

2017-12-9 First Release

See *oreilly.com/catalog/errata.csp?isbn=9781680452969* for release details.

978-1-168-045296-9

Safari® Books Online

Safari Books Online is an on-demand digital library that delivers expert content in both book and video form from the world's leading authors in technology and business. Technology professionals, software developers, web designers, and business and creative professionals use Safari Books Online as their primary resource for research, problem solving, learning, and certification training. Safari Books Online offers a range of plans and pricing for enterprise, government, education, and individuals. Members have access to thousands of books, training videos, and prepublication manuscripts in one fully searchable database from publishers like O'Reilly Media, Prentice Hall Professional, Addison-Wesley Professional, Microsoft Press, Sams, Que, Peachpit Press, Focal Press, Cisco Press, John Wiley & Sons, Syngress, Morgan Kaufmann, IBM Redbooks, Packt, Adobe Press, FT Press, Apress, Manning, New Riders, McGraw-Hill, Jones & Bartlett, Course Technology, and hundreds more. For more information about Safari Books Online, please visit us online.

How to Contact Us

Please address comments and questions to the publisher:

Maker Media, Inc.
1700 Montgomery Street, Suite 240
San Francisco, CA 94111

You can send comments and questions to us by email at *books@makermedia.com*.

Maker Media unites, inspires, informs, and entertains a growing community of resourceful people who undertake amazing projects in their backyards, basements, and garages. Maker Media celebrates your right to tweak, hack, and bend any Technology to your will. The Maker Media audience continues to be a growing culture and community that believes in bettering ourselves, our environment, our educational system—our entire world. This is much more than an audience, it's a worldwide movement that Maker Media is leading. We call it the Maker Movement.

To learn more about *Make:* visit us at *makezine.com*. You can learn more about the company at the following websites:

Maker Media: *makermedia.com*
Maker Faire: *makerfaire.com*
Maker Shed: *makershed.com*
Maker Share: *makershare.com*

Contents

Acknowledgments

Love and gratitude to my daughter, Annika, for making my life a whirlwind of discovery and creativity, and to my awesome partner, Jodi Kittle, for helping me find focus and balance in the midst of that chaos. Thanks to Shelly Willie, who invited me to Chadwick International, which in turn offered the incredible pleasure of daily collaboration with Gary Donahue, whose innovative spirit shines through in every mBot project. Sylvia Martinez's and Gary Stager's Constructing Modern Knowledge Press not only connected me to the powerful history of children and computers, but also the incredible cohort of Maker-educators whom I'm lucky enough to call colleagues and friends, including Josh Burker, Jaymes Dec, Angi Chau, Karen Blumberg, and Brian C. Smith.

Andrew

As in all my life, my wife, Angie, and kids, Kelly and Micah, provide constant fun, encouragement, and inspiration. I love you guys so much! I'm thankful to my dad, Bill, for allowing me to work alongside him as a kid and learn along the way. I so much appreciate my staff team and the awesome students and parents at Steindorf K–8 STEAM School in San Jose, California, where I teach. Starting a new public school has been a wild ride, especially while writing a book! Thanks to Andrew for his willingness to share his expertise and endure endless questions with a chill attitude. As a lifelong Maker, I have been given so many opportunities by Maker Media over the past ten years to do what I love doing—Making and teaching. Thanks!

Rick

About the Authors

RICK SCHERTLE has taught middle school for over 20 years, and now runs the Maker Lab at Steindorf K–8 STEAM School in San Jose, California. Rick has been involved in Maker Faire for many years. He has written nearly two dozen articles for *Make: Magazine*, including his first article in volume 15 in 2008 on compressed air rockets. He also wrote the book *Planes, Gliders, and Paper Rockets* from Maker Media. Rick is the cofounder of *AirRocketWorks.com*.

ANDREW CARLE has taught in K–12 schools for 15 years. He launched the Makers program in 2010 while teaching programming and math at Flint Hill School in Northern Virginia. In 2014, he moved to Korea to expand Chadwick International's school-wide Making & Design program. He has presented at Maker Faires and has with MakerEd.org, National Association of Independent Schools (NAIS), Virginia Society for Technology in Education (VSTE), and International Society for Technology in Education (ISTE), and has been named a Senior FabLearn Fellow for Stanford's Transformative Learning Technologies Lab.

Introduction

The Arduino came to prominence as a tool to help designers, artists, and musicians access the power of inexpensive Atmel 8-bit microcontrollers. The Arduino allowed people with deep skills in another discipline to bring their ideas to life. All it took was learning "a little programming" without having to acquire the full range of skills to work with embedded electronics.

That mission has been so successful over the last decade that it created a need for a new tool, one that could connect young people with minimal skills to the "little programming" world of Arduino.

Over the last five years, there's been an explosion of kid-friendly programming and robotics tools. After working with dozens of different kits and boards, we became deeply impressed with the mBot. But as the technology choices multiplied, the tutorials and introductory materials offered didn't match the ways we used these platforms in classrooms, Makerspaces, and clubs. We saw the need for a book that offered instructions for specific projects, in conjunction with advice on using the mBots with large groups in a classroom setting. The mBot allows novices to start with idle tinkering on the base mBot, and access higher-level features or add new components when inspiration strikes. This flexibility is crucial for classrooms or cohort groups, since the mBot allows raw beginners and experienced tinkerers to work at their comfort level.

Shenzen, China–based Makeblock has emerged as a major player in the kid-focused robot kit market. Their mBot is the cheapest and most widely available—you can buy them on Amazon and directly from their website at *www.makeblock.com*—with hundreds of thousands of mBots distributed around the world. While the mBot kit and the many accessories available for the mBot are well engineered and made from quality materials, there is a lack of technical support and documentation. The Makeblock website has an active forum and user base, but information is often confusing and hard to find. We hope to bridge the divide between a quality product and the thousands of users with this first-of-its-kind book.

Rick and Andrew met during the summer of 2016 as coaches for the Design Do Discover (D3) conference at Castilleja School in Palo Alto, led by mutual friend Angie Chau.

Rick was an initial supporter of Makeblock's mBot Kickstarter campaign. He bought five mBots for a neighborhood Makerspace he was helping start. Due to the affordability of the platform, several years later he purchased 20 more mBots for the new school where he would be running the K–8 Maker Lab. Those mBots would become a core part of his Maker Lab curriculum.

Andrew had come to D3 with several deconstructed mBots, and was using the mCore controller by itself for a variety of creative uses. In South Korea where Andrew taught, he had over a hundred mBot kits that were used at nearly every grade level of his K–8 school.

Rick's desire to learn more about the platform and the ability to scale it up for classroom and school use was a perfect match with Andrew's firsthand experience. Facebook Messenger conversations began across time zones, and this book is the result of this cross-Pacific collaboration. We hope this book cuts through much of the quirks and confusion resulting from the mBot documentation, saves you time when scaling up your mBots for classroom and school use, and gives you some creative project ideas to use right away.

1

Kit to Classroom

Right out of the box, the mBot has many features appealing to kids and adults alike. But the true power comes from the heart of the mBot—the Arduino-based mCore microcontroller, and the other sensors and actuators in the Makeblock platform. These components transform the retail mBot kit from a Christmas morning diversion into a classroom powerhouse, and offer plenty of possibilities for anyone who's "done everything" with the mBot.

OUT-OF-THE-BOX KIT

When people think of mBots, what usually comes to mind is a cute little robot with an anodized aluminum body and sturdy components. You can buy mBots at all major retailers online and in brick-and-mortar stores.

The mBot chassis, wheels, motors, controller, and sensors can be put together in about half an hour with the screwdriver (included). Even young kids can follow along with the clear, IKEA-esque visual instructions included in the commercial kit. While that flyer is helpful, we're going to walk through the steps of putting together your mBot in a bit more detail.

FIGURE 1-1: The mBot

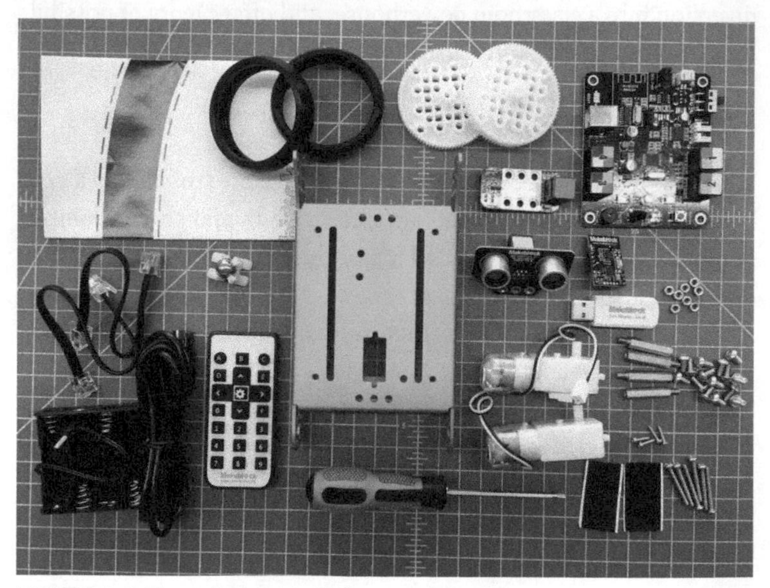

FIGURE 1-2: The mBot parts right out of the box

INSTALLING THE MOTORS AND WHEELS ON THE CHASSIS

We'll begin mBot assembly with the sturdy anodized aluminum chassis. Gather the following parts and let the building begin!

Parts

Chassis, M3×25 bolt (4)	Wheels (2)
M3 nuts (4)	M2.2×9 screw
Motors (2)	Tires (2)

Quick Primer on Simplified Sizing

In the parts list, you can see part names for bolts and screws with seemingly a scramble of letters and numbers. Let's unscramble the meaning.

- The *M* in *M3x25* stands for *metric*. Metric screw threads were one of the first globally standardized parts established by the International Standards Organization in 1947.
- The number to the right of the M is the diameter of the threaded part of the bolt.
- The number after the × is the length of the bolt measured in millimeters
- The bolt shown in Figures 1-3 and 1-4 is an M4x8. It's 4 mm diameter (Figure 1-3) and 8 mm long (Figure 1-4).
- Metric bolts can use either hex head wrenches or Phillips screwdrivers.

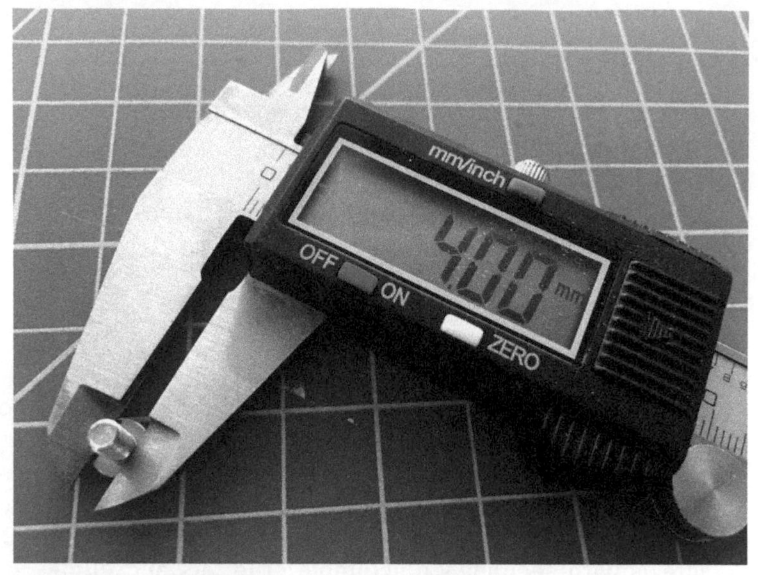

FIGURE 1-3: Measuring the diameter of a bolt

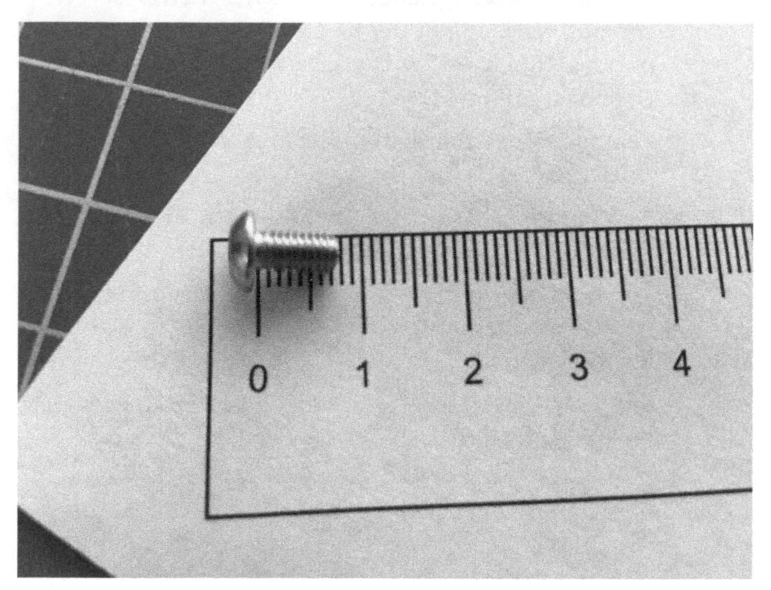

FIGURE 1-4: Measuring the length of a bolt

Variety packs of metric bolts, nuts, and washers (like the one shown in Figure 1-5) can be purchased on Amazon for a good price. Many of the projects in this book will use these small metric bolts in a variety of lengths. Makeblock's large aluminum beams and building materials all use M4 bolts. The mBot chassis uses M3 bolts and spacers, along with smaller M2.2 bolts to affix the wheels.

FIGURE 1-5: Cornucopia of hardware

Steps

1. Line up the holes on the geared motor with the holes on the chassis, and insert the M3×25 bolts. Tighten a M3 nut on the end, holding it in place while tightening to prevent it from spinning.

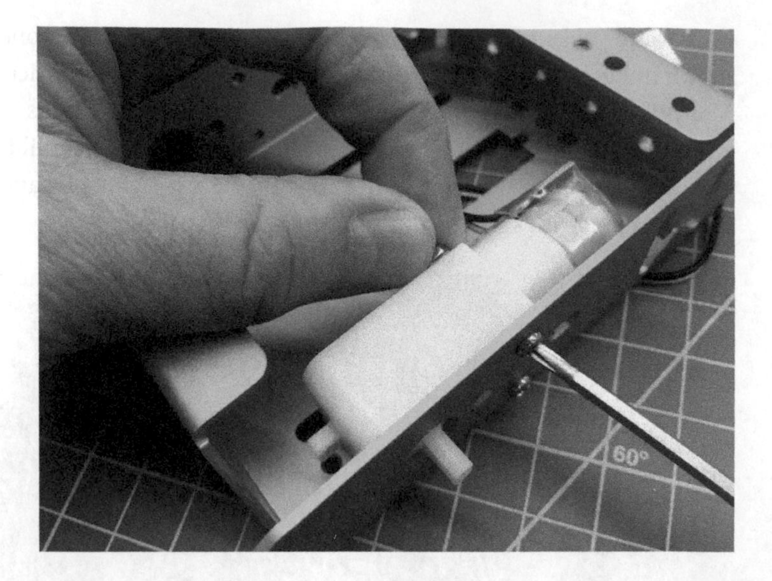

2. Repeat with the second motor. The motors are identical, so it doesn't matter which side you install them on.

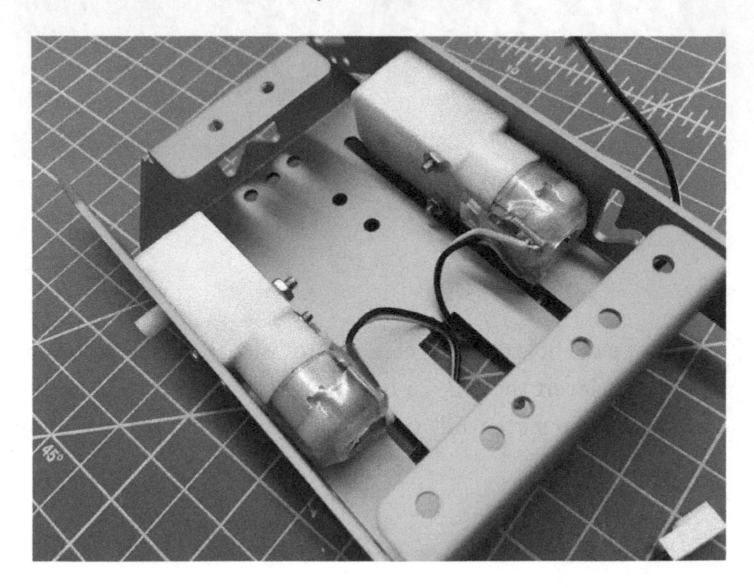

3. Feed the motor wires through the top of the chassis.

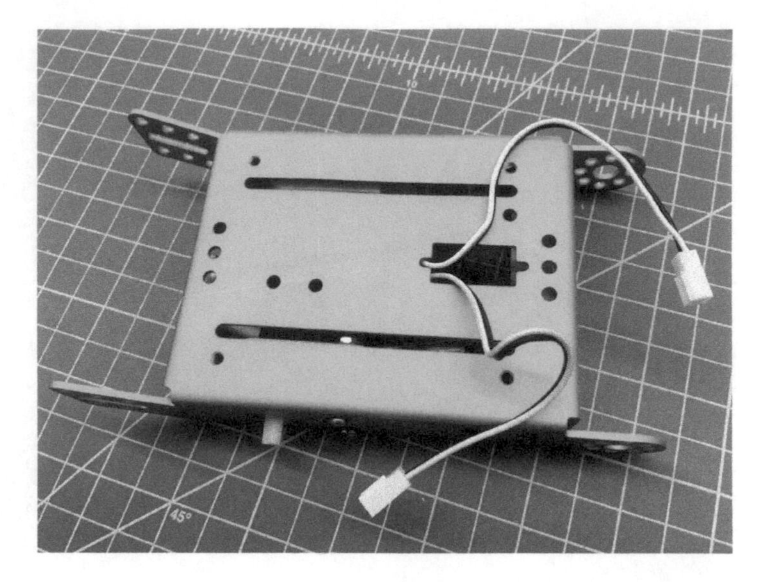

4. Using an M2.2×9 screw, attach the wheel to the geared motor. It's easy to strip the Phillips head screw by overtightening, so tighten the screw only until it's snug, and then stop!

5. Attach the tires over the wheels.

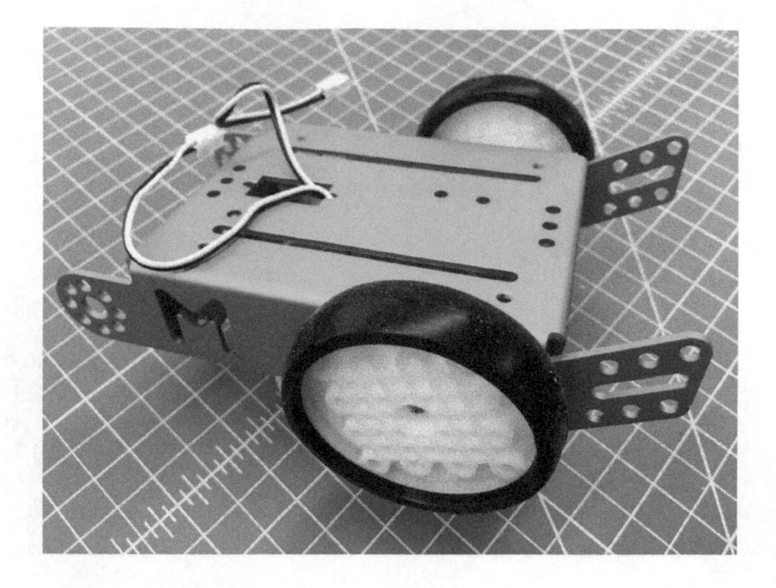

About the Motors and Wheels

The mBot's aluminum frame is designed to fit the geared DC motors included in the kit. The design is pretty standard across the robotics world, but you may find small inconsistencies in hole spacing or shaft depth between third-party motors and the mBot frame.

Makeblock includes a few replacements for the internal plastic gears with the motor. If you're opening and building a bunch of kits at once, be sure to grab them before kids turn them into tiny tops.

The mCore board only has one power circuit, which provides regulated 5V power to the microcontroller and the motor ports via an onboard H bridge. This means that the M1 and M2 motor ports are limited to DC motors or pumps that operate at 5V. Unlike other robot platforms or Arduino boards, the mBot can't connect a second, larger power source that would supply power to just the motor ports. If you need more oomph than the included geared motors can provide, Chapter 6, "Building Big and Small with mCore," shows how to expand the mBot using external power relays.

In later chapters, we'll cover how you can add LEGO gears and pulleys to your wheel hubs to power all kinds of things with your motors.

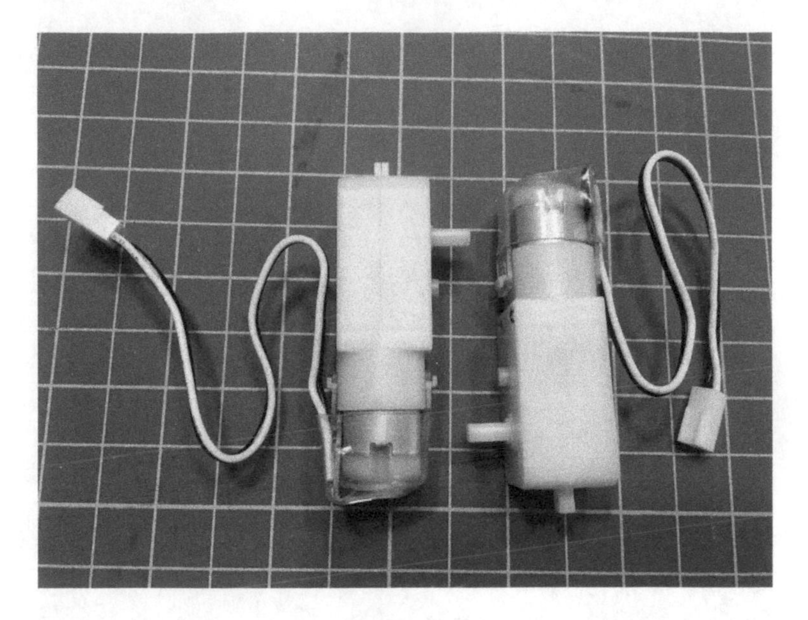

About the mBot Chassis

Makeblock's first product was a set of anodized aluminum construction materials, which included rectangular beams and open and threaded connection points. These beams are still at the core of many Makeblock products, including the XY Plotter and 3D printers.

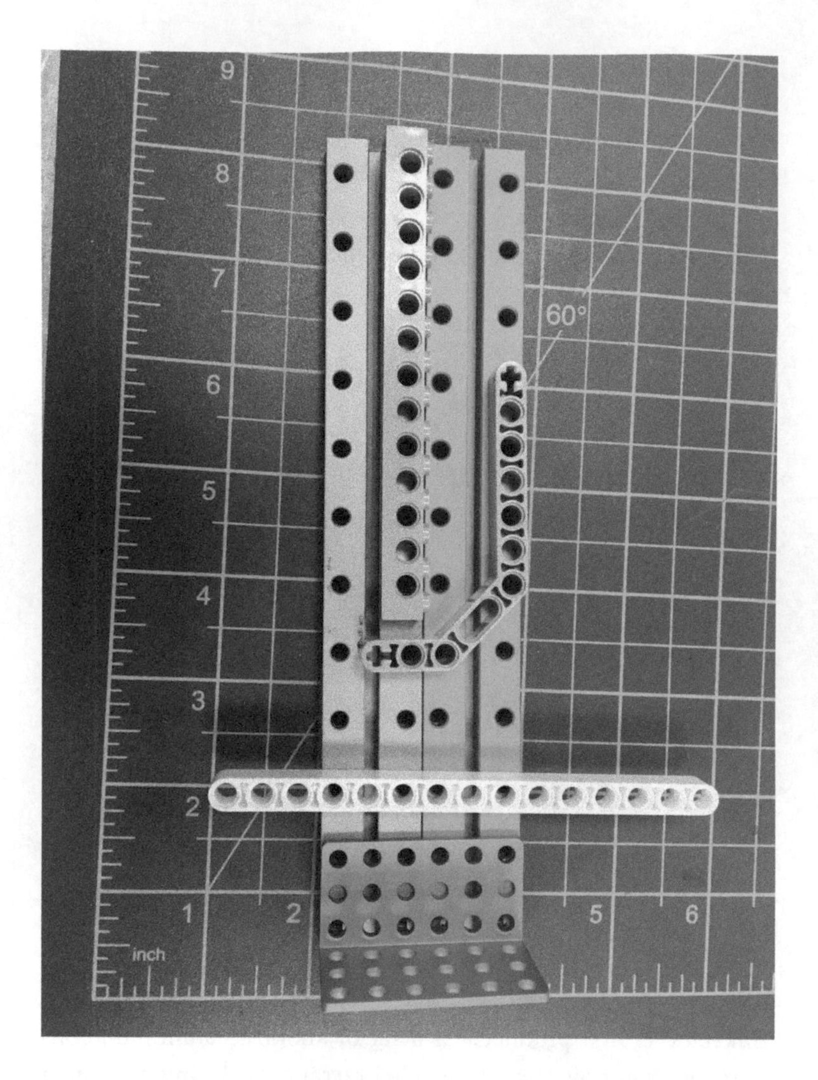

In the early days, Makebock was very clear that their aluminum parts were designed to work well with LEGO Technic parts. Holes on Makeblock parts are spaced to neatly overlap with the most common Technic beams, although the holes are sized slightly smaller. Makeblock holes are sized for 4 mm metric screws. To fit these screws, Makeblock holes are just a hair over 4 mm, while Technic holes are 4.8 mm. This size discrepancy means that although LEGO Technic pins cannot connect a Makeblock beam to LEGO parts, standard M4 bolts and nuts can anchor LEGO to Makeblock parts.

Over the years, Makeblock's product line has expanded, and now there are many examples of products on both sides that can't connect easily. Consequently, Makeblock doesn't advertise the physical compatibility between their hardware and LEGO Technic anymore. However, the M4 sizing remains consistent across all sorts of strange and specialized parts. The aluminum mBot chassis continues this tradition.

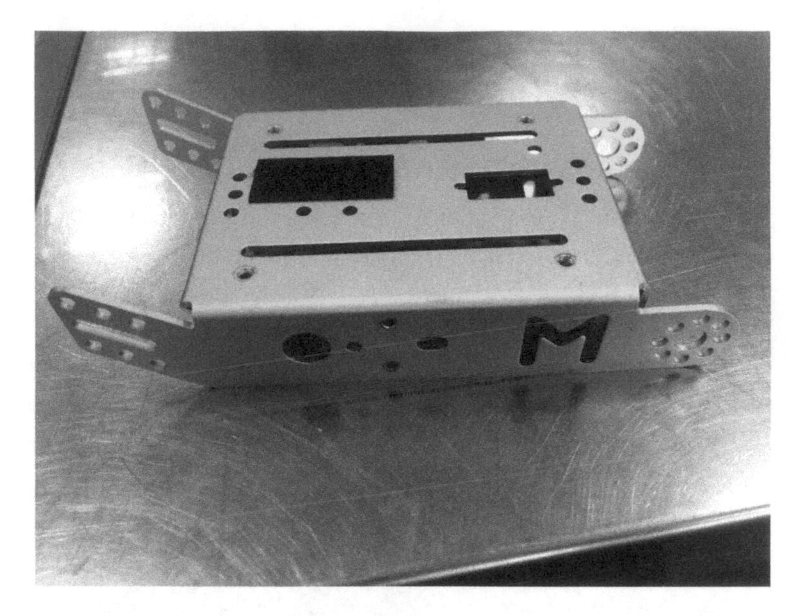

The chassis is covered with M4 holes, most often spaced apart to fit Technic, to provide multiple anchor points and easy expansion. There are a few mounting points on the mBot frame that are smaller than M4, notably the two points where the yellow motors attach, and the larger holes for the motor axle. When working with the mBot and LEGO, you'll need M4-14 screws or longer. An M4-14 screw will go through one Technic beam and the thin aluminum on the circular or angled tabs, with room for one nut. Thicker parts will naturally require longer screws. When working with mBots and LEGO in the classroom, we keep a collection of M4 screws handy, with lengths ranging from 15 to 40 in 5 mm increments.

INSTALLING THE SENSORS

Next, we're going to install the sensors. Let's get started with the parts you'll need.

Parts

Me Ultrasonic Sensor	6P6C RJ25 cables (2)
Me Line Follower	M4×8 screws (4)

Steps

1. Flip your mBot chassis upside down. On the front underside of the chassis, line up the middle holes on the line-following sensor with the roller ball stacked on top.
2. These holes in the chassis are threaded, so you just need to screw both M4×8 screws down through the holes in the roller ball.

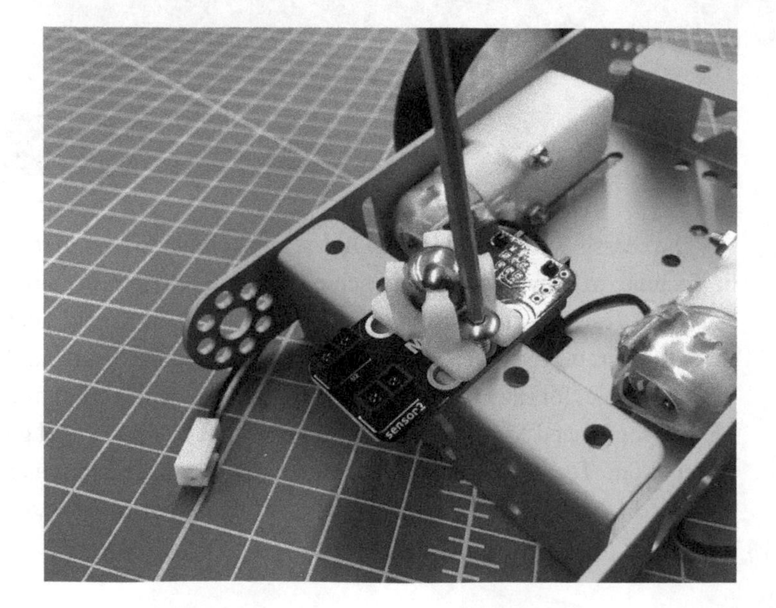

NOTE The shaft on the screwdriver pulls out and has a Phillips tip on one end and a hex tip on the other.

3. Flip the mBot back over, and line up the holes on the ultrasonic sensor with the holes on the front of the mBot above the "smile." Screw these together with two M4×8 screws.

4. Plug the RJ25 cable into the line-following sensor, and feed the wire through the opening in the chassis. Plug the other RJ25 cable into the ultrasonic sensor.

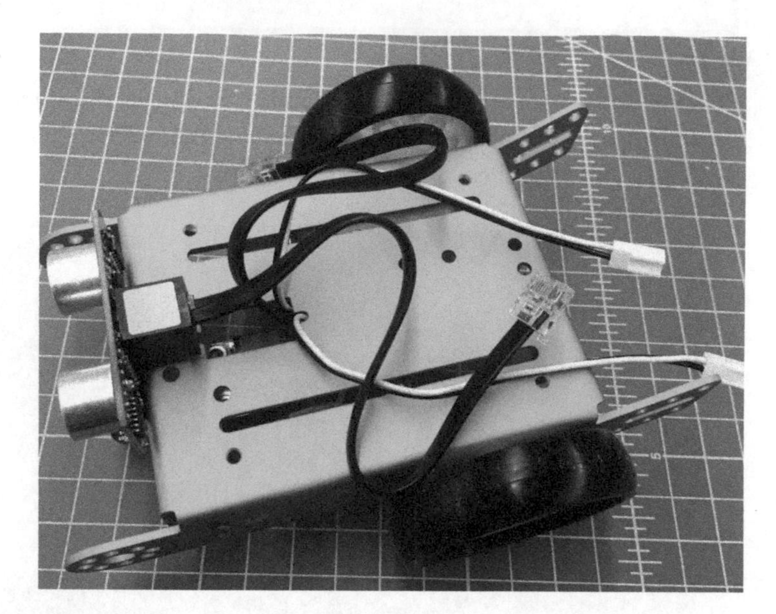

> **NOTE** A full assortment of available add-on sensors are described in the table at the end of Chapter 2, which includes pictures and descriptions of each sensor and sample code for testing the various sensors.

ADDING THE BATTERY HOLDER

Parts

4 AA battery holder

Step

Connect the barrel jack on the 4 AA battery holder to the DC power jack on the mCore as shown.

INSTALLING THE MCORE AND BATTERY

Parts

mCore board	M4×8 screws (4)
mBot chassis	5 cm of Velcro
M4×25 brass stand-offs (4)	

Steps

1. Screw the brass stand-offs into the four pre-threaded holes on the top of the chassis.

2. Lay down a strip of the included Velcro on the back of the chassis.

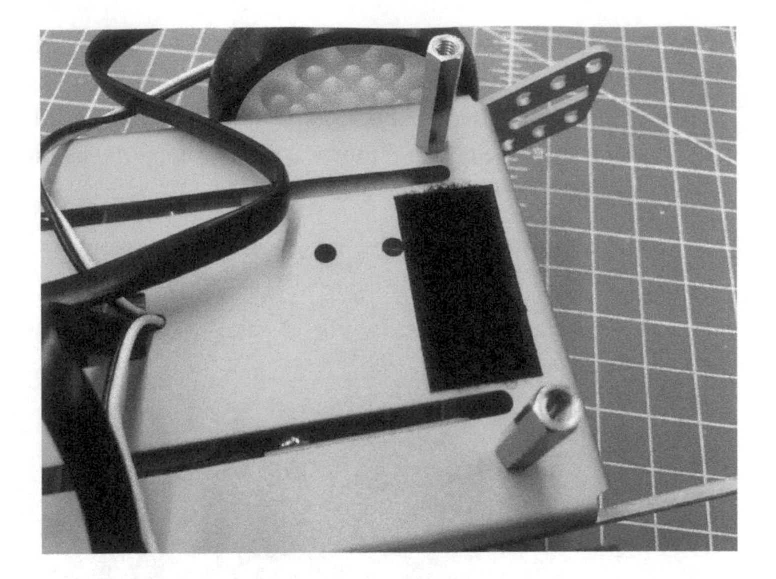

3. Attach the other half of the Velcro to the back of the 4 AA battery holder. The power cable for the battery pack should point out toward the back.

4. Stick the battery pack down onto the chassis, as shown. Now run the wires so the ultrasonic sensor cable lies along the right side of the mBot (when the back of the mBot is facing you). The line-following cable and two motor wires lay along the left side of the mBot.

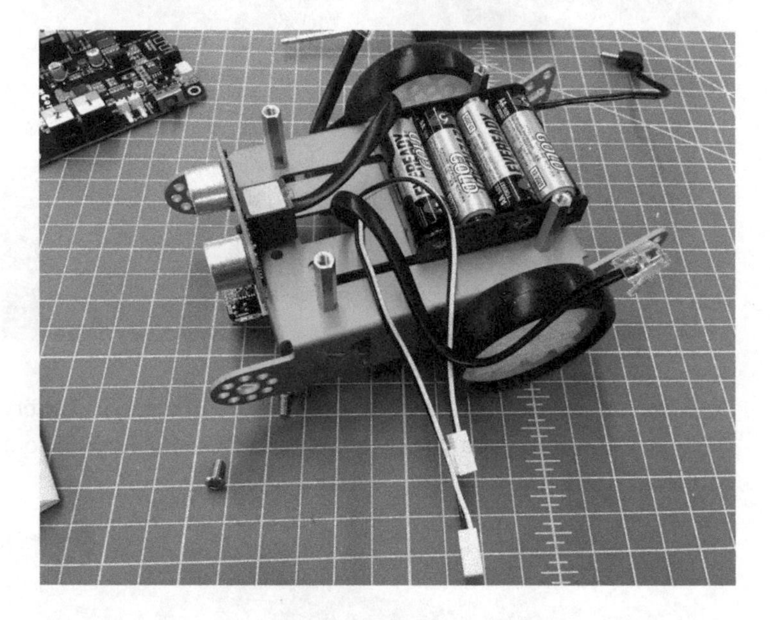

5. Place the mCore down on the brass standoffs, as shown, and secure with M4×8 screws.

> **NOTE** A detailed tour of the mCore board takes place in the "Classroom" section later in this chapter.

WIRING THE MBOT

Parts

Everything is already installed. Just follow these steps to connect correctly.

Steps

1. Looking at the mCore from the top, plug the ultrasonic sensor into port 3.

2. Plug the line-following sensor into port 2.

3. Plug the motors into the two motor jacks on the left side of the mCore.

4. The barrel plug on the battery can be plugged into the round power jack on the back of the mCore.

COMMUNICATING WITH YOUR MBOT

Parts

Bluetooth or 2.4G module

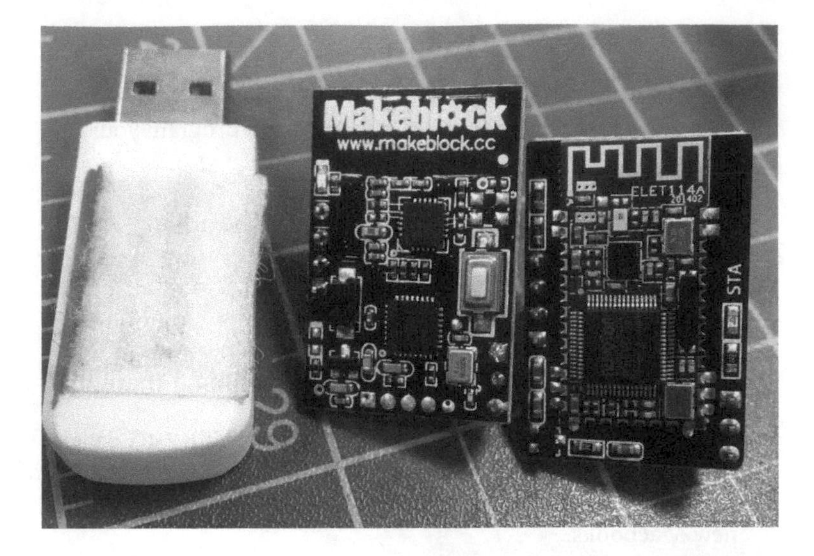

Both versions of the mBot come with an infrared remote (IR); however, when buying an mBot, you need to **specify either the 2.4G or Bluetooth version.** In the accompanying image, the Bluetooth module is on the right, and the 2.4G (GHz) module (with its USB dongle) is on the left. Either module plugs into the mCore board in the wireless module slot on the left rear corner.

Makeblock's advertising copy distinguishes these models by referring to them as School (2.4 GHz) and Family (Bluetooth), a useful if imprecise summary. This confused me when I was buying my first mBots.

Let's take a look at the differences between the two.

Pros and Cons of Bluetooth

- It's easy to connect to a Bluetooth-enabled tablet or laptop computer. By doing this, you can control and program your mBot using Bluetooth.
- It's the best option when working with just one mBot.
- It takes more work to pair and the process is platform-dependent.

Pros and Cons of 2.4G

- It's easy to connect to any computer only using a USB 2.4G dongle.
- Requires a "classic" USB port or a USB-C adapter to use with new Macbooks.
- This is by far the preferred option when using many mBots with a group of kids. Kids can just plug in the dongle, connect, and begin programming!

In order to connect the module, just line up the four pins on one side and three pins on the other, and insert the Bluetooth or 2.4G module into the slots on the mCore—that's it!

TEST THE MCORE FOR CORRECT CONNECTIONS

Flip the power switch and you'll hear three tones. The two front left and right lights (LED1/2) will flash red/green/blue, then off. A red power light in the middle of the circuit board (PWR) stays on, along with another red light (very small) on the back of the range sensor. Two tiny blue lights on either side of the line-tracking sensor should also stay on when your mBot is placed on the table or if you put your finger over them. The line-following sensor also has a tiny red power light.

> **NOTE** If one or more of these lights is not lit, check the connections on ports 2 and 3 and check the batteries.

TEST YOUR MBOT REMOTE

Insert a CR2025 battery into the remote, making sure the battery is installed with the smooth + side facing toward the remote buttons. The remote only has about a four-foot range and requires line of sight to the IR receiver on the front of the mCore. There are three modes preprogrammed into the mBot or mCore to use with the remote: modes A, B, and C.

MODE A: REMOTE MANUAL CONTROL

When you select this mode, you'll hear a low-tone beep, and the two LEDs on the front of the mCore will turn white. In manual control mode, the arrows on the remote control the direction of the robot, and the numbers adjust the speed of the robot, with 1 being the slowest and 9 the fastest. If any of the buttons don't work, check the motor connections and make sure the batteries are good. Try pressing 9 (full power) and try the other buttons again using the higher power level. If left and right turn in the wrong direction, the motor wires may be reversed. If the wheels aren't turning, check to make sure all wires are plugged in and that the battery has a full charge.

> ## Which Is the Left Motor?
>
> The left motor is the one installed under connectors 1 and 2. Both motors are the same, but once they're installed, they become *left* and *right*. The left motor should be connected to the white power plug beside connector 1.

MODE B: WALL AVOIDANCE/RANGE CHECKER

When you select this mode, you'll hear a medium beep, and the LEDs will turn green. To see it in action, hold the mBot in the air and press B. The wheels will turn. As you move your hand in front of the range sensor, the wheels will change direction for a moment and then return to normal. If this does not occur, the range sensor may not be connected. Check to make sure the red power light on the back is lit. Ensure that the range sensor is connected to port 3 on the mCore, which is the only port that will work for the demo program. Make sure it's snapped all the way into the sensor as well.

MODE C: LINE-FOLLOWING

When you select this mode, you'll hear a high-tone beep, and the LEDs will turn blue. To see how this mode works, open the folded sheet with the giant number 8 on it, and place the mBot right on top of a black line. Turn the mBot on, and press C. The mBot should immediately start following the black line, adjusting its wheels to follow the line as it moves. If this does not happen, confirm there are two blue power lights on the tracking sensors. Make sure the tracking sensor is plugged into port 2.

WHAT TO DO WITH YOUR MBOT RIGHT OUT OF THE BOX

Now it's time to get creative and artsy with your standard mBot. Many materials (craft sticks, cardboard, straws, and so on) can be added to the front and rear racks of the mBot frame by either bolting them on with M4 bolts and nuts, or using a hot glue gun.

> **NOTE** If you're going to attach things with a hot glue gun, put masking tape on the frame first so the glue will come off without damaging the frame.

The following image shows a neat idea for building a rack for the front and rear of an mBot.

PROJECTS

Although the mBot is a powerful and programmable robotics platform, there's a lot to explore using just the mBot's IR remote. In this section, we'll explore activities you can start the moment you tighten the last screw on the mBot chassis. These are great for opening meetings when lots of folks are assembling mBots at once. Nothing motivates you to finish the last of the wiring like the chance to join a pick-up game of robot soccer.

In this section, we'll look at some cool things you can do with the basic mBot setup we just finished.

Here are some activities you can do with just the IR remote (included in both the Bluetooth and 2.4G kits):

- Race around a DIY obstacle course—go ahead and set up some cups on the floor, and make ramps, and so on!
- Run timed races through the obstacle courses.
- You could create a fancier obstacle course by requiring the use of the three preprogrammed modes:
  - First, steer around cones using Mode A.
  - Second, find the black line and begin line-following in Mode C.
  - Third, switch to Mode B, obstacle avoidance, to get through a maze.
- Attach a pen or pens to the front or rear of the mBot to turn it into a drawing bot.
- Make parades with multiple mBots using Mode C, the line-following feature. Chapter 2, "mBot Software Sensors," has more information on how this works, including how to add sensors to make them navigate autonomously.
- Move a load of straws or blocks from point A to B (providing different parameters for different age groups) using racks built onto the front and back of the mBots. With younger students,

if the robot simply moves with a load, this might count as success, whereas older students might need to navigate bridges or tunnels moving both forward and in reverse. If you're using multiple mBots, teams can be timed for a competition.

» Create an extension to the mBot that moves some object to perform a task; for example, you could add an iPad to create a mini telepresence robot, or add floor scrubbers and sweepers.

With mBots that are paired to a computer or tablet using Bluetooth, several (or many) mBots could be controlled independently. Here are some ideas you can try using the Bluetooth module:

» Sumo wrestling—Draw a big circle on the ground with tape, and the mBots can try to push the other bots out of the ring.

» BattleBots—Attach a BBQ skewer to the front of the mBot and a balloon to the back. The mBots must try to pop each other's balloons. Learn more about this in Chapter 2.

» Race course—Race head-to-head through an obstacle course the kids build.

TO CLASSROOM

During the last six years, there's been an explosion in boards, kits, and tools roughly described as "kid electronics." In that time, I've used (almost) all of them in my classroom. Although a few of those products became MakerEd workhorses, most failed in serious ways when put into the hands of real students in a classroom Makerspace. I was looking for a low-floor, high-ceiling open platform that allows students to start with their Scratch programming skills and transition out into "real" Arduino.

> **NOTE** Scratch is a free graphical programming language developed by the Lifelong Kindergarten Group at MIT. With millions of users, it's a familiar and accessible tool for everyone from kids to adults. Scratch can be used to program a variety of Arduino-based microcontrollers.

Makeblock's mCore board is the microcontroller that powers the mBot, and it comes as close to the classroom robotics bulls-eye as any other product available. Although the board was created and released as part of the mBot kit, it's now available directly from Makeblock at a significantly lower price. Even without the chassis and motors that ship with the kit, the mCore is a great learning platform.

The mCore board uses an Atmel ATmega328, common across many boards of the Arduino Uno generation. Instead of the traditional Arduino shield layout, many of the digital and analog I/O pins are routed into the four phone jack plugs. Several basic components are built into the board, including some RGB LEDs, a buzzer (outputs), a push button, and a light sensor (inputs).

ONBOARD COMPONENTS

Makeblock electronic components use a 6-pin "phone" plug (known as RJ25 or 6P6C). The components and ports are color-coded so that components that require specific features from the AT328 will always be matched to the right pins. There is a great chart to illustrate this at the following website: *http://learn.makeblock.com/makeblock-orion/*. (See colored square shapes labeled 1–4.)

WHITE

This is the serial port for I2C devices. Many devices in the Arduino universe use a serial protocol called I2C. Devices with existing Arduino libraries can be used with the mCore in Arduino mode. However, there's currently no way to access I2C devices through the mBlock programming interface.

BLUE

Makeblock refers to components that go in this port as *double digital*, which simply means that the sensor sends or receives data over both digital I/O pins. Some of the other Makeblock boards have ports without blue, but all four of the mBot ports can be used for double digital.

YELLOW

Devices that go here all use a single digital I/O port.

GRAY

This is the hardware serial—none of the four ports on the mCore have the gray label, because the RX/TX pins run to the wireless module.

BLACK

Components that require analog input ports Arduino pins A0–A3 belong in this port. Examples include any sensor that reports a variable resistance, like a potentiometer (slide, knob, or analog stick). The mBot has black connectors on ports 3 and 4 only.

RED

While there are no red ports on the mBot, other Makeblock products use red for motor ports that tap into a higher voltage line (basically, Vin for the Arduino). The mBot does not have a secondary power supply on the main board, so it doesn't need a red port. We cover the different ways to use larger motors with the mBot in Chapter 6.

On the mCore, all four numbered ports have white, blue, and yellow markings. This means they can use any of the digital sensors or 12c devices. Only ports 3 and 4 also have black, so the mCore is limited to only two simultaneous analog sensors.

If you're interested, the specific Arduino pin number that corresponds to each plug is silk-screened onto the board behind the RJ25 plug.

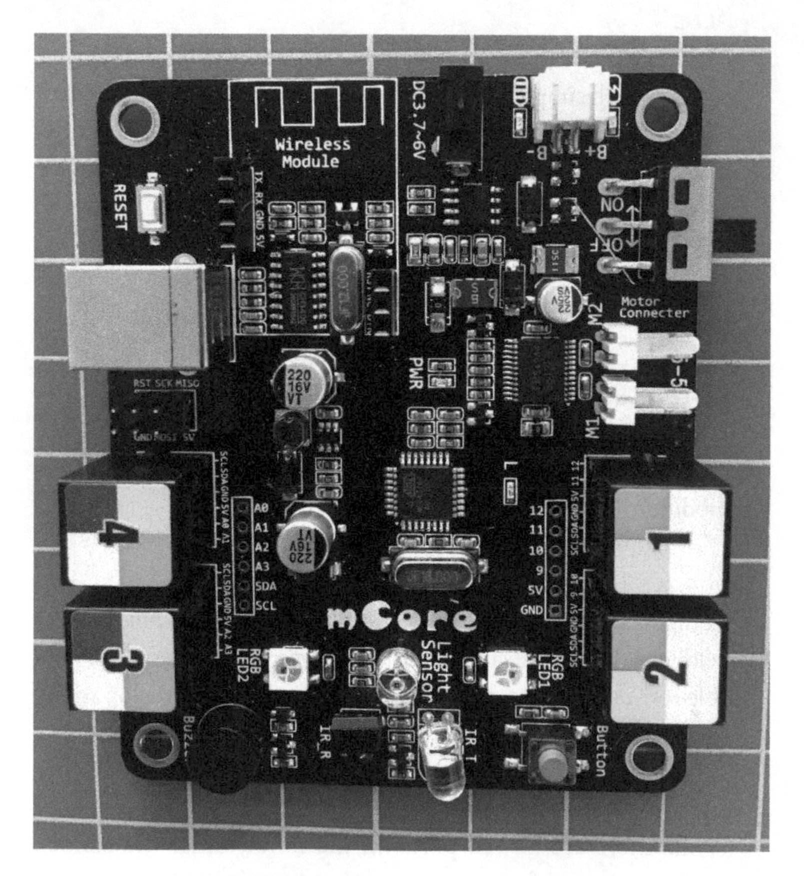

POWERING UP YOUR MBOT

There are three plugs that can accept a power source for the mBot: USB, the 2.5 mm barrel plug, and the two-pin JST lithium ion battery (LIB) connector.

USB is probably the most familiar option with new users. The connection on the mBot board uses the hefty USB-B plug, normally seen on printers and other large devices. When compared to the USB micro or mini used on other Arduino-inspired boards, the USB-B plug is downright burly. This weight and stability is a huge benefit when working with kids. While USB can obviously be used for data, it works just fine as a simple power port. Using a short USB A or B cable, you can power an mBot from a standard external USB battery for many hours. Note that supplying power to the USB port does not activate the board unless you also turn the power switch on. It sounds obvious, but that's different than normal Arduino boards.

The mBot ships with a 4 AA battery holder that uses the 2.5 mm barrel plug. This plug is smaller than the standard Arduino 3.5 mm barrel plug, possibly to serve as a last-minute reminder that it is *not* safe to power an mBot with a 9V battery.

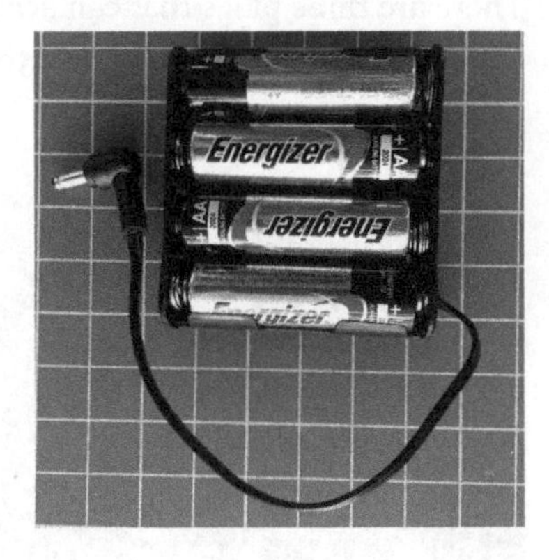

The JST connector is a mixed blessing for classroom use. Once it's docked, the connection is incredibly snug (yay!), to the point where kids who attempt to unplug the battery will often rip wires out of the harness (boo). If rechargeable batteries needed to be removed and reattached on a daily basis, the JST connector wouldn't survive a month. Thankfully, the mCore includes an onboard charging circuit, so that LIBs connected to the JST port can charge when the mCore is connected to a power source. You'll need to provide power through the USB port or the barrel plug to charge an attached LIB. When charging the LIB over USB, treat it like any other rechargeable electronics. While you can charge them one at a time off of a computer, it's best to use a dependable 1–2A USB charger. Being able to charge five mBots from a good quality USB charger hub is a lifesaver when working with classroom sets.

PWR
M2
DC3.7~6V
TX RX GND 5V

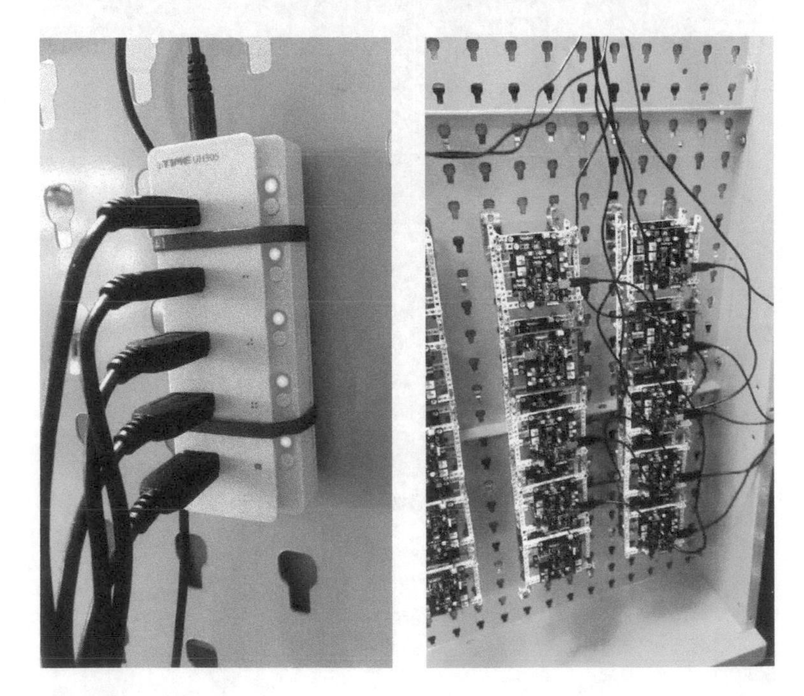

TOUR OF THE MCORE AND ONBOARD SENSORS

The mCore includes a few basic components on the board itself. These don't constitute a full sensor suite, but they're components that support simple behaviors on the default (car-like) mBot platform. Chapter 2 has a chart with the onboard sensors with MBlock Scratch code to test them.

We'll go over the components of the board, starting at the bottom right of the mCore board, and moving up.

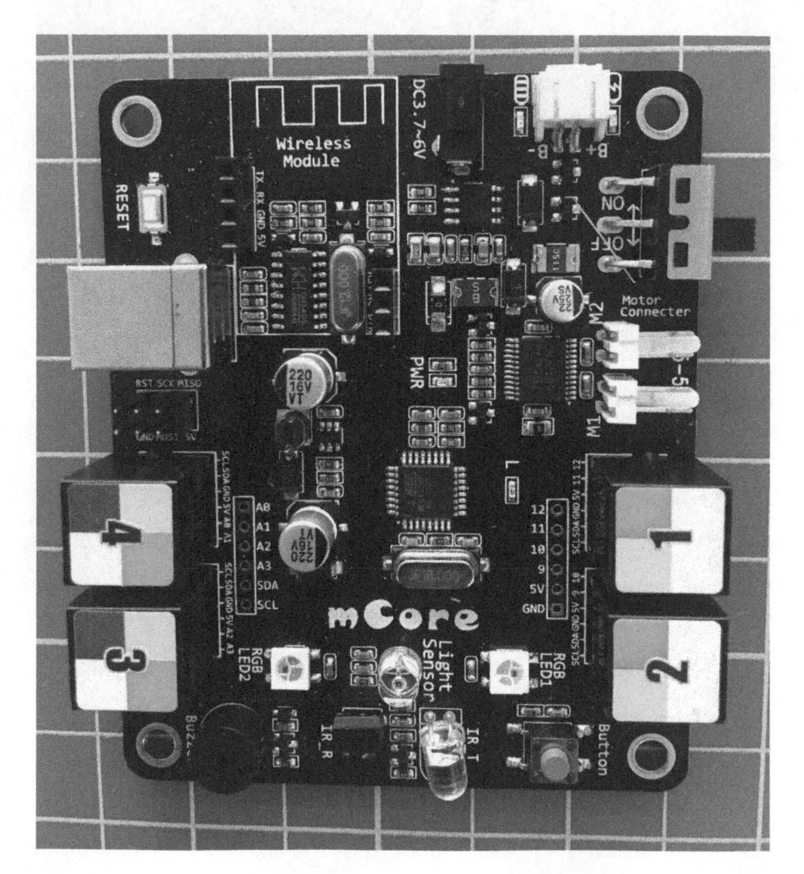

There's a simple push button in the bottom-right corner of the board. It's not fancy, but it's useful for programs where the mBot needs to be put into position before the wheels start turning.

Next to the push button, there's an infrared receiver and transmitter. With the default program loaded on the mCore, the receiver is set up to move in response to commands from the included IR remote. Every mBot and remote is set up the same way, so commands from any remote will affect all mBots in range. This is great for semi-synchronized hordes of roaming robots, but really frustrating for kids who want to play robot soccer against each other.

In the bottom-left corner, there's a piezo buzzer. Pleasing those with an '80s nostalgia for abstract bloops and squeaks, Makeblock distinguishes between their different programs with small tones or chirps when the board starts up. This seems cute rather than crucial, but losing track of which board holds which program can create huge headaches in a classroom setting. Imagine you're staring at a table full of mBots and knowing that *one* of them has a student-created program loaded. Without a Makeblock program loaded, the mCore will fail to connect to any programming environment, but will not provide any clear error message. Knowing that the boards with the correct program make a distinctive sound allows you to check that a table full of mCore boards is ready for use in under a minute. Thank you for your service, humble buzzer.

There are two programmable RGB LEDs in the second row. These LEDs are mounted in series and use a single signal wire to control a tiny (seriously; super tiny!) microcontroller built into the plastic housing, which then passes instructions down to the next light. There are only two lights in this series on the mCore board itself, but the same type of lights are used on the Makeblock LED board and longer LED strips.

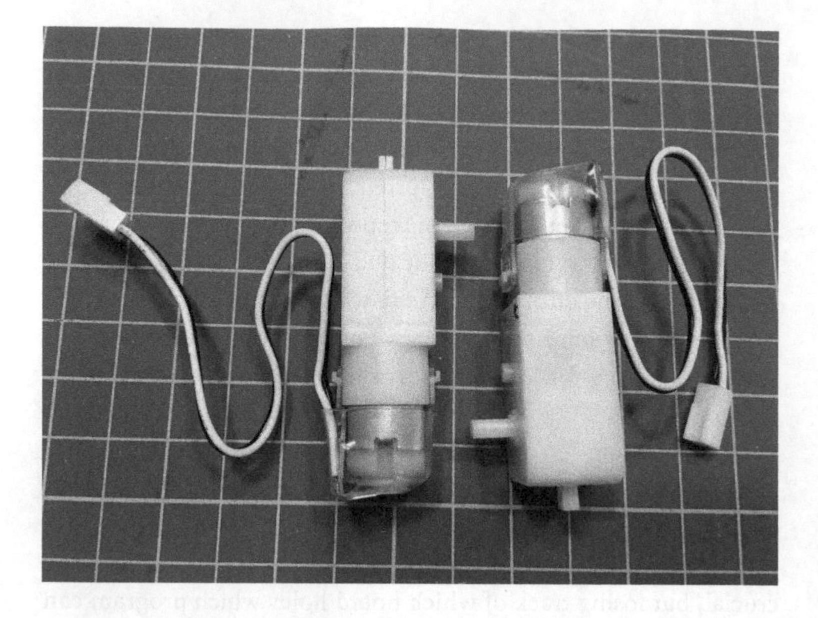

The onboard sensors just described in detail are built right into the mCore—the brains of the mBot. The add-on sensors listed in the table at the end of Chapter 2 are available for purchase individually and in bundled packs for very reasonable prices. One of the strengths of the mBot platform is that the price for standard components in Makeblock packaging isn't astronomically high. Nearly all the add-on sensors can be connected to the mCore using RJ25 (phone jack) cables. For sensors that are not made by Makeblock, the RJ25 adaptor is the perfect solution.

Every parent has a story about the surprising amount of damage kids can instantly inflict on small electronics. Teachers have even more stories, and theirs include mysterious damage or loss to components over school breaks, when the school is supposed to be locked. When non-educators visit our Makerspace, my colleague, Gary Donahue, reminds them how much chaos one kid with a bucket of LEGOs can unleash, and asks them to extrapolate that out to 30, 60, or 120 kids working with materials in a given day. Even 10 kids with LEGOs can thrash your living room, and they'll transform a carefully curated set of LEGOs into a fully homogeneous mess.

The quiet challenge of a robotics program in a club or school setting is making sure kids have access to the same materials on week two as they do on week 26, and ensuring that the room can reset quickly after each session.

STORING COMPONENTS

There are two basic schools of thought regarding the storage of small components in lab or classroom settings: by kit or by kind.

Kits are great for large homogeneous exercises, where each group will tackle roughly the same problem with the same materials. Makeblock sensors and motors are small enough that plastic pencil cases make great storage containers. Small kits can also help younger kids learn organization and cleanup skills. Even if the parts jumble around inside the container, a color-coded inventory on the inside lid really helps the end-of-class inventory.

In other settings, simply grouping the same types of parts into accessible bins may work better. On our physical computing carts, Makeblock parts are grouped into motors, lights, servos, simple sensors, complex sensors (compass and gyro), and external motor boards. Louvered bins make it easy to set up a cart for classroom use with all the parts we would include in a kit. Although this does make it easier to miss an individual piece during clean-up, it also drastically reduces the number of components out on student desks at any one time. When all the parts are sitting right there on the rack, students will (with some encouragement) walk up and grab materials only when needed.

STORING PROJECTS

Nothing kills a robotics project faster than bad storage. As an individual, maybe you can claim an entire table for the duration of a project. In a club or classroom setting where everything has to be put away and ready for another group several times a day, that's never an option. It's crucial to think about how you're going to store both materials and in-process projects. A great storage solution will both minimize the disruption caused by cleaning up the work area, and ensure that everything is ready to go next time.

Storing Basic mBot Projects

The mBot comes in a very nice little cardboard box that stacks well (see Figure 1-6). For any class or project where students are exclusively programming the basic robot, and not adding structure or sensors, I'm happy to keep using those boxes for project storage.

When students are programming the mCore boards, there's little incentive to even assign particular robots to groups of kids. For a programming project that uses the standard mBot vehicle design, different groups of kids can use the same robot all day long. Programs made using a tablet or sent from mBlock using Bluetooth or wireless aren't actually written to the internal memory on the mCore.

FIGURE 1-6: The sturdy cardboard box mBots ship in

Instead, the programming environment on the tablet or computer sends instructions over the wireless connection. When you reset the mCore, the default program loads up and is ready for the next batch of kids. All of the important stuff is stored on the tablet or computer. With the addition of a good lithium polymer (LiPo) or LIB, one set of mBots can support classes all day long.

Storing the Assembled mBot

Once students are making additions or modifications to the mBot, the cardboard box is no longer a good option. Not only is it too cramped for kid-made stuff, but when there's a variety of sensors and parts in use, being able to survey those parts at a glance is crucial.

Throughout our Makerspace, we use heavy, broad stacking tubs for in-process project storage. Choosing a single, standard bin has many quiet and unexpected benefits for classroom organization.

But it's not always possible to devote that amount of space to individual projects. For years, we had enough rugged LEGO bins to use for this purpose, but we eventually outgrew them. The closest match we found (because secondhand LEGO bins are staggeringly expensive!) are IKEA TROFAST bins, which have a similar footprint and low sides. A low, wide bin like this makes it easy to run charging cables to each robot, and to easily put parts in and take parts out, even when the bins are on shelves.

NOTE One way to help each mBot kit serve more kids is to use a standard, easily removed frame for attaching sensors and actuators. If the additional parts move as a unit—sensors clip on, sensors clip off—then kids can remove their additions at the end of each session and leave a clean mBot for the next group. This only adds a few minutes to clean-up procedures and allows one set of mBots to serve a whole grade, or even a whole school. There are instructions and templates for different frames included in the downloadable resources for this book.

Storing an mCore with Mixed Materials

Once the cardboard and popsicle sticks come out, and your mBot is much bigger and more complex than the standard factory bot, a good storage plan is critical. When working with large groups, a visible, consistent storage container can define maximum size for a project without any explicit instructions. We like durable bins that come in a few different heights while keeping a consistent footprint, like the IKEA TROFAST line.

If the mCore boards are going to stay in student bins, students must ensure that the USB port stays accessible for charging. Since the mCore's motor and sensor ports are so close to the USB plug, students will normally need to keep that area accessible throughout the project.

Sensors that use RJ25 plugs and motors can be connected and disconnected quickly. If you have long cables, it's reasonable to ask students to build their sensors and motors into a structure that can cleanly detach from the mCore board. This arrangement allows builders to detach the specialized (and cheaper) parts of their work from the (more expensive) mCore at the end of every session. Then the mCore units can return to the charging system when the project bins return to the shelf.

PROTECTING THE MCORE

I have a few recurring nightmares around kids and electronics: among the worst, baskets of parts dropped down the stairs, and components left on the floor and stepped on or crushed beneath casters. LEGO has a well-established position at the top of the kid-safe electronics pyramid, meaning their electronics are safe *from* kids. The assembled mBot isn't quite that stable, but there are many ways you can improve its odds of survival.

The most vulnerable part of the mCore board by far is the wireless module slot where the Bluetooth or 2.4G serial boards attach. When a board is mounted in this slot, it sticks up slightly higher than the USB-B plug and, during free-fall, has an instant attraction to the floor.

The best way to protect a component is to reduce or eliminate reasons for students to touch it. If you're considering a frame or case for the mCore, work hard to ensure that users clearly understand how and where to hold it, and inspect the frame to make sure those areas are far away from the weak points.

Starting with the v1.1 mBot kit, Makeblock now provides a semitransparent plastic case that mounts directly through the board to the brass standoffs. These work well if you are using the standard robot with wheels. But many of the projects throughout this book use the mCore as a stationary computing platform instead of as a robot. In those cases, it makes sense to remove the mCore from the robot chassis. Without that bulky aluminum frame, there's no way to attach the v1.1 mBot cover.

USING A LEGO TECHNIC FRAME

Every workshop or classroom has unique needs, and the best solution should meet those needs exactly. When working with the standard vehicle created from the mBot kit, we found the chassis and components stable and kid-resilient. The gap between the mCore board and the chassis is large enough to fit the 6 AA battery pack or a large, rechargeable lithium battery. After a few days of kid use, we added a small strip of Velcro between the battery pack and the aluminum frame to secure the battery when the kids were carrying the robot around.

The aluminum frame provides excellent stability and protection for the mBot, but it's also bulky. Many projects in our Makerspace use the mCore as a physical computing platform that doesn't need to move, or one that needs better ways to connect to LEGO, cardboard, or other craft materials. For those projects, it's a real hassle to work with the assembled mBot on the aluminum chassis. But without the stability provided by that frame, it was clear that the bare mCore would need something to hold the battery in place and prevent strain on the JST plug.

Driven by that initial need, we developed a basic frame from LEGO Technic. Much of the experimentation and iteration came from our colleague Gary Donahue at Chadwick International, who was always looking for a way to trim just a few more LEGO blocks from the design. Instead of simply protecting the bare mCore board while in

FIGURE 1-7: This rolling rack stores and charges up to 50 mCore boards off a single wall outlet.

use, this frame simplifies the logistical challenges that come from large groups of people working with the mCore. This design lifts the board off the table, provides connection points for LEGO or Makeblock parts, and preserves access to the USB and sensor ports. When you're working with class sets of mCore, you need to easily charge 20 or more boards on a cart that can move from room to room, while using as few LEGO beams as possible. This frame (shown in Figure 1-7) represents our current solution.

Check here for the pieces you'll need to make the LEGO Technic frame: *www.airrocketworks.com/instructions/make-Mbots*.

The following image shows all the pieces laid out with nylon nuts and bolts and hex nuts and bolts.

The next image shows what the finished frame will look like. This frame will make it easier to store and protect your mCore. It will also hold the battery underneath.

Although it's not mentioned in the mBot materials, the corner holes for the brass standoffs are perfectly aligned to LEGO Technic spacing. This frame takes advantage of that fact to provide a support structure that holds and protects an LIB, maintains easy access to the RJ25 and USB ports, and lies flat on a table.

The first two beams are attached directly under the mCore board, aligned with the sensor and battery sides. Using a 15-hole LEGO beam, insert an M4 bolt through the corner hole nearest the button on the mCore, through hole 4 of the beam, and then close with an M4 nut. Repeat that process, putting a bolt through the hole nearest the buzzer and hole 12 in the LEGO beam. Then, add a second 15-hole LEGO beam using the holes next to the battery port and the Reset button on the mCore, as shown in the following image.

Secure underneath with an M4 nut.

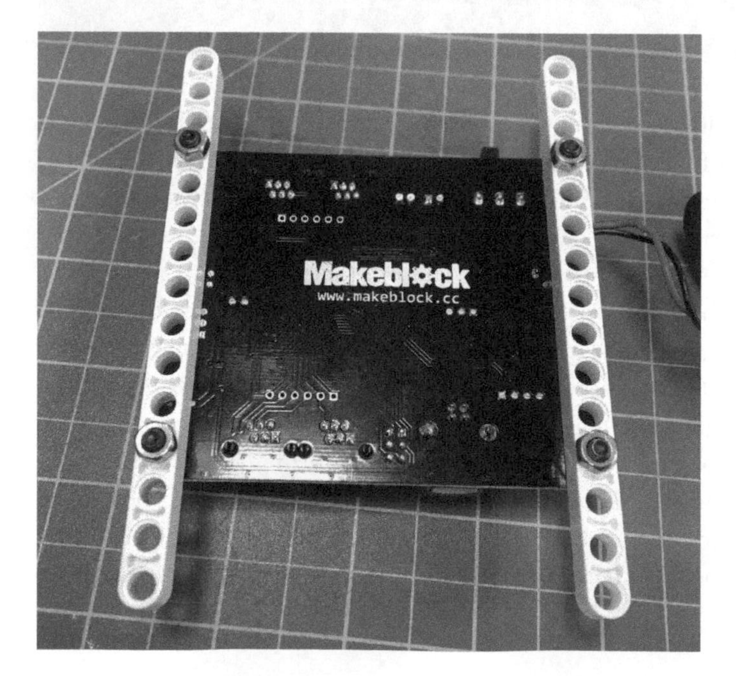

With the mCore still flipped over, insert a LEGO long pin with friction into the ends of the 15-hole beams with the long ends going through the beam.

Attach the other 15-hole beams to the LEGO friction pins, as shown next.

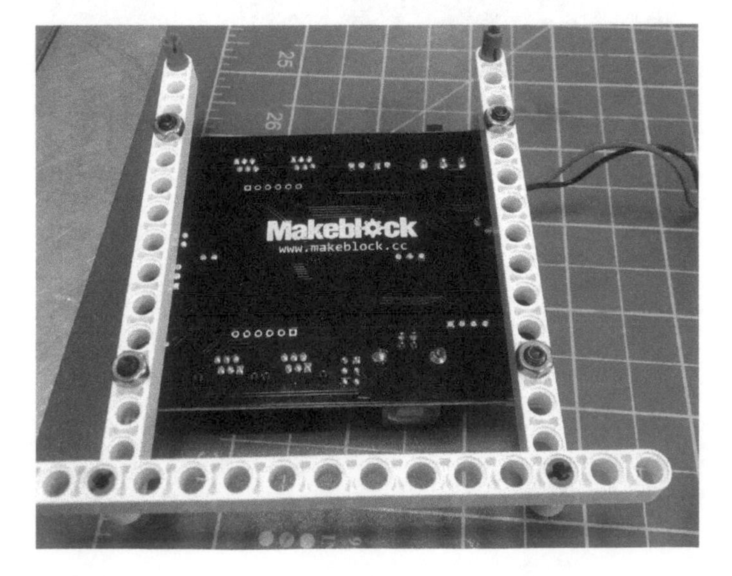

Now, place a longer M4 bolt through hole 8 in the end beams and secure with a nut. Set your battery holder on the bottom of the mCore, as shown. This M4 nut also serves as a spacer, to provide clearance over the LIB holder. Be sure to check your battery holder's size and add or remove spacers, as necessary.

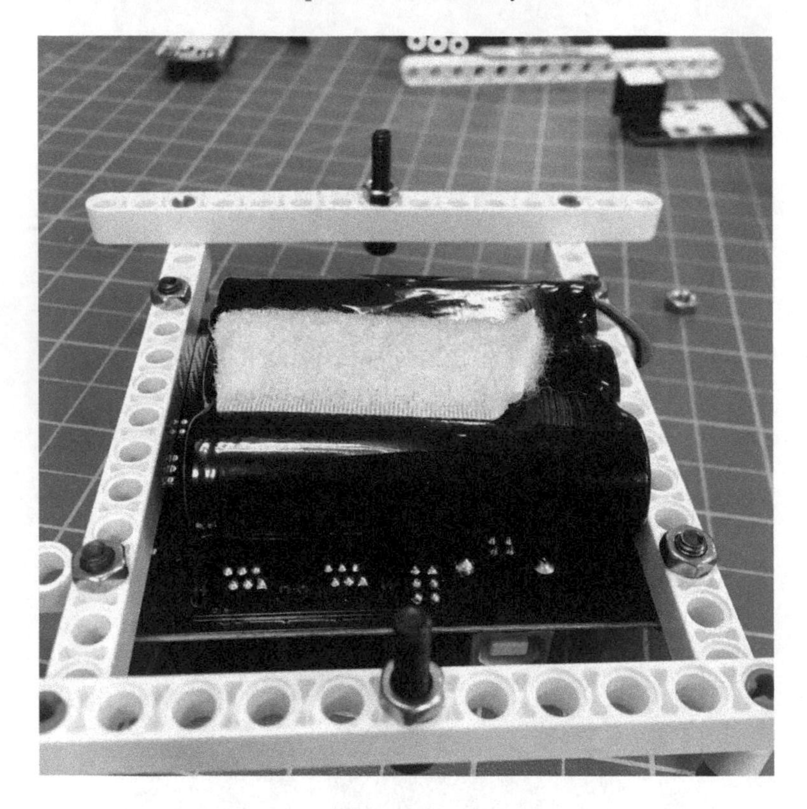

Now, add a final LEGO 15-hole beam over the bolts and screw it down using an M4 nylon or steel nut. This will keep the battery holder firmly in place, without compressing the battery cells.

The following images show what it looks like finished! The photo on the right shows the 2.4G USB dongle attached to the battery with Velcro so it doesn't get lost. Also, add two LEGO Technic Cross Blocks 1×3 to one side of the frame for hanging many mCore's on a rack. Sweet!

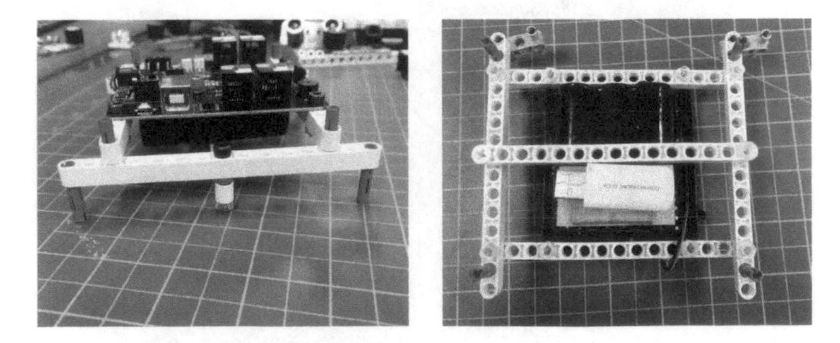

Once you get the Technic frames done, you can easily hang many of them from a frame for easy storage (see Figures 1-8 and 1-9).

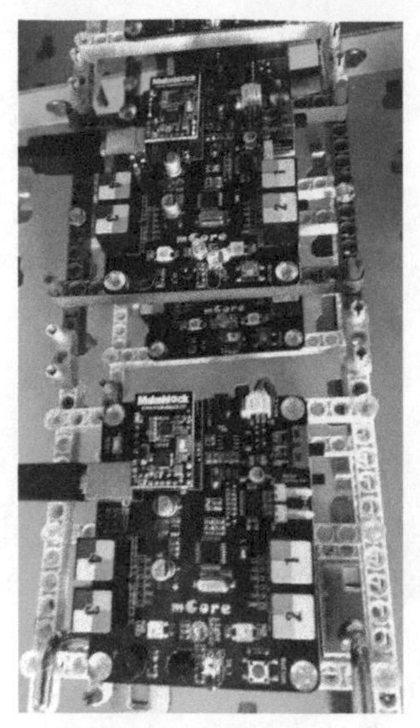

FIGURE 1-8: Here is a close-up of the mCores hanging on the rack.

FIGURE 1-9: Rick's version of the LEGO Technic frame

ADDING A COVER

Using the M4 brass standoffs from the mBot frame assembly, you can easily add a simple cover. If you have access to a laser cutter, or a drill press and patience, you can expose the lights and light sensors, and even extra points of connection with a few rows of Technic-spaced holes.

Here is a custom, laser-cut cover I designed, made from ⅛″ acrylic. This cover protects the 2.4G serial or Bluetooth connection and has five holes on each side for LEGO connection points or connections with other Makeblock add-on pieces. Laser cut files are available at this book's website: *www.airrocketworks.com/instructions/make-mBots*. If you don't have access to a laser cutter, full-scale files are also available in PDF form for hand-cutting.

Storage for the mCore is underneath the batteries. Attach battery holder to the bottom of the mCore using Velcro with adhesive. Install the battery holder so batteries face the smooth-bottomed acrylic pieces to avoid the chance of the batteries shorting out against the bottom of the mCore.

While these frames and cases (and the others available on the book's resource page) are useful, they might not meet the specific needs of your program. Experiment with the materials you have on hand until you develop a cover, frame, or storage system that fits your classroom perfectly! (See the cool DIY case in Figure 1-10.) Then share it back with us!

FIGURE 1-10: Case made from 32 oz, 4″ × 4″ Ziplock container by "John1" on the Makeblock forums

MAKING CABLES

Cabling is often at the heart of proprietary control schemes. Everyone who lived through the digital camera explosion probably has a drawer full of USB cables with weird, manufacturer-specific ends. In the educational robot sector, cabling is what transforms standard servos, motors, and sensors into premium branded components.

Makeblock does use a standard connector on the mBot, but the type is not obvious from first inspection. The RJ25 connector looks like a standard United States phone plug, but it's a specific version of that standard. Makeblock uses a 6P6C *modular jack*, meaning that it has six contact points connected to six actual wires.

Making your own cables for this plug requires a crimping tool. Most Ethernet crimping tools have ports for the smaller modular plug, as well as the larger 8P8C plug used for Category 5 or 6 wire. Although you can use twisted pair Ethernet wire for mBot cables, I find unwinding the pairs to be a huge hassle. Using flat six-wire cable makes the process swift and easy.

Parts

Six-wire cable—often you can find this cheaper and in reasonable lengths when sold as single long phone cable, rather than as a bulk cable package. One 100ft phone cable will create many classroom-sized connections.

6P6C/RJ25 modular plugs—make sure these have six-wire contacts, not 4.

Crimping tool—most label the connection we need as *RJ11/RJ12* or *Phone*.

It is important to keep the color alignment consistent between the two ends of the cable. When inspecting a plug, you should see the same color order on the wires going from left to right. With the wire shown here, white is on the left-hand side of the plug and blue is at the far right. Although these colors may vary by cable manufacturer, they need to be consistent between ends of an individual cable.

Working with cable ends is another version of the "my left, your right" problem, where changes to the orientation of the parts makes relative direction useless. This perfectly useless tiny cable shows that plugs put on either end in the same orientation will reverse the order of the wires between the two ends. (See Figure 1-11.)

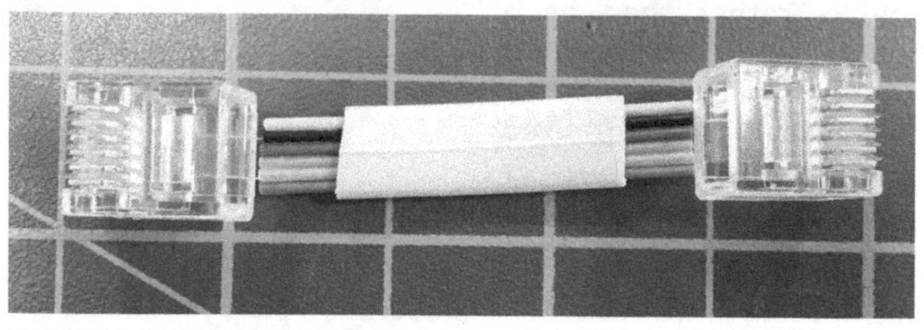

FIGURE 1-11: This is the *wrong* way to crimp a Makeblock cable!

The plug on our left will have a left-to-right pin order of white-black-red-green-yellow-blue. The right-hand plug will have a left-to-right pin order of blue-yellow-green-red-black-white. Maintaining color order will result in wires where the plugs are rotated 180 degrees from each other—especially noticeable on small cables. (See Figure 1-12.)

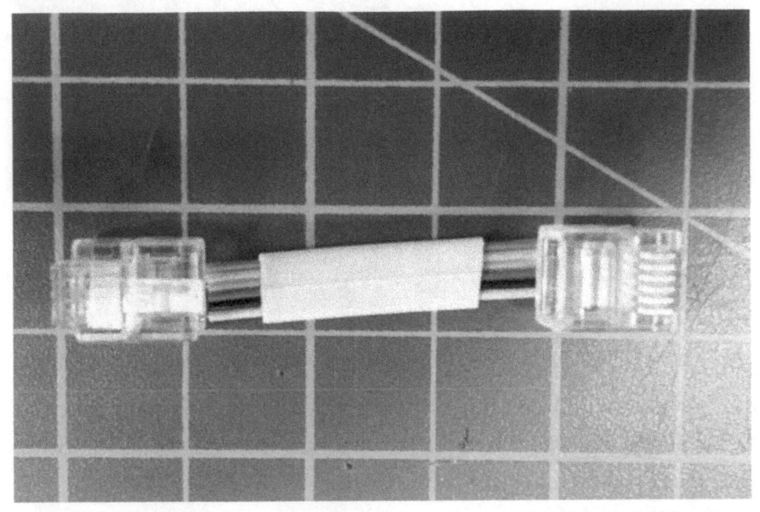

FIGURE 1-12: This is the correct way to crimp a Makeblock cable. To keep wire order the same, the ends must be reversed.

Each time I make a cable, I slide the plastic ends on with the metal prongs facing me and recite the colors in order. If the colors match at both ends, the cable will be fine.

Now, it's time to gather the materials and build some cable.

Steps

1. First, cut the desired length of cable from the spool and then strip about 1 cm of housing from each end. When using modular connectors, you do *not* need to strip the individual wires.

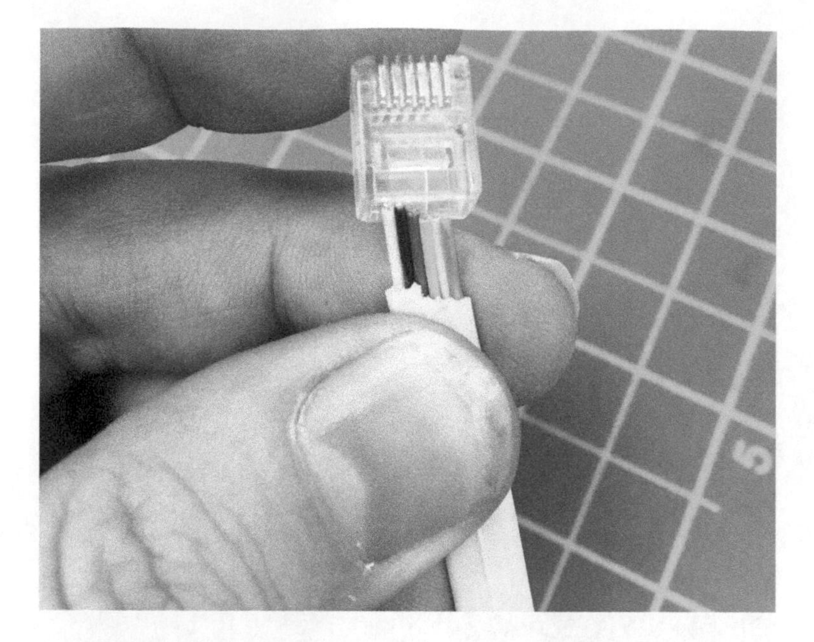

2. With the outer coating stripped, slide the modular jack over the exposed, colored wires. Ensure that all six wires slide smoothly under the metal prongs of the jack.

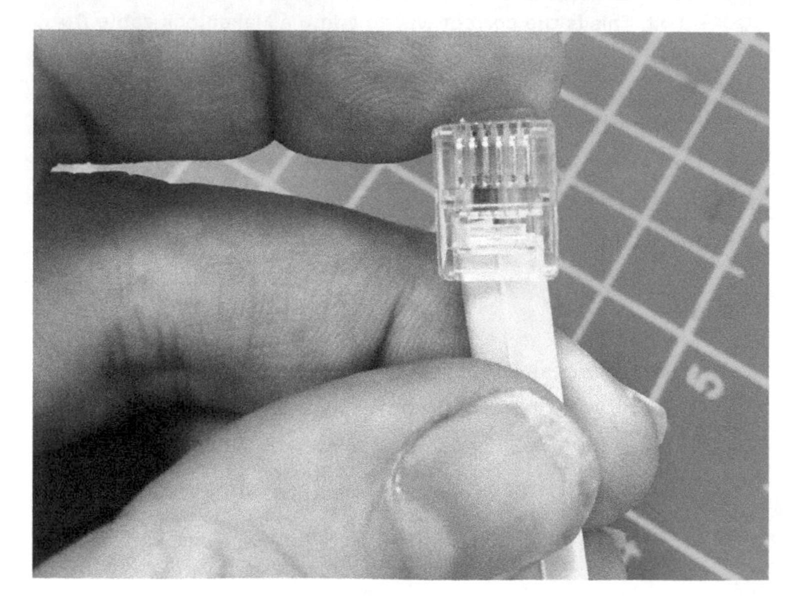

3. Before you crimp, look directly at the end of the jack. You should see the cross section of all six colored wires at the same depth, underneath the brass teeth of the plug. If one wire is shorter than the others, it will appear further back and less distinct. To save yourself some headaches later, remove the plug and re-trim the wires so that they're all flush, then replace the plug and check again. Missing the connection on one wire out of six invites a world of inconsistent and intermittent errors, depending on which wires a particular add-on uses to communicate with the mBot.

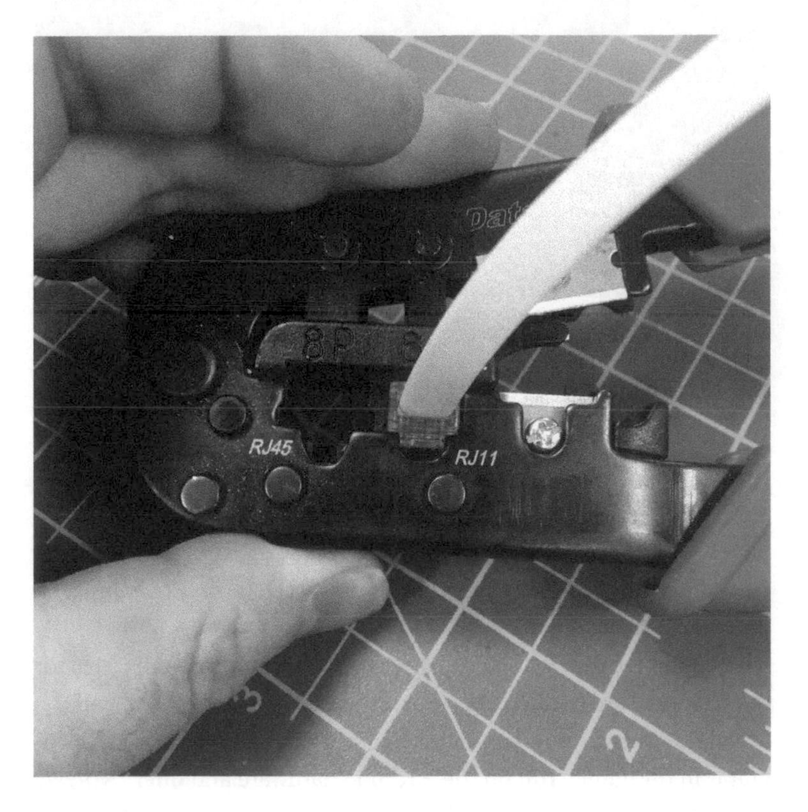

4. Place the cable and plug into the crimping tool, then squeeze. It doesn't take much force to drive the metal pins into the colored wires. Check one last time to make sure you can see all of the metal teeth biting into to each of the six wires.

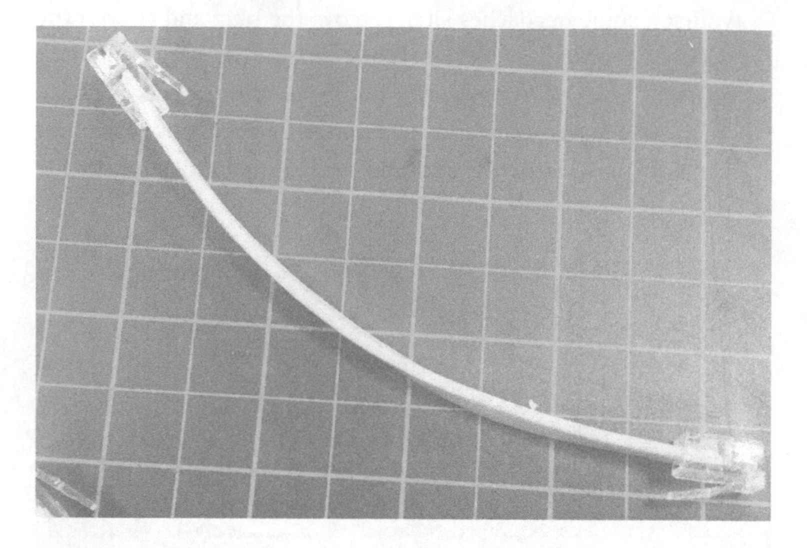

UPDATING THE MBOT

Within mBlock, there are two different pieces of software that can run on the mCore board and connect to mBlock or the Makeblock app. While they both appear in the mBlock Connect menu, the labels leave a lot to be desired. One is labeled Update Firmware (see Figure 1-13), and the other is Reset Default Program (see Figure 1-14).

Despite the different names, these are both Arduino programs for the mCore board based on the open source Firmata protocol and the StandardFirmata program developed over the last decade. All programs in this family run on the Arduino hardware and offer two-way communication between the physical board and a computer. That task eats up much of the limited program memory on the mCore, leaving little room for extra mBot-specific functions.

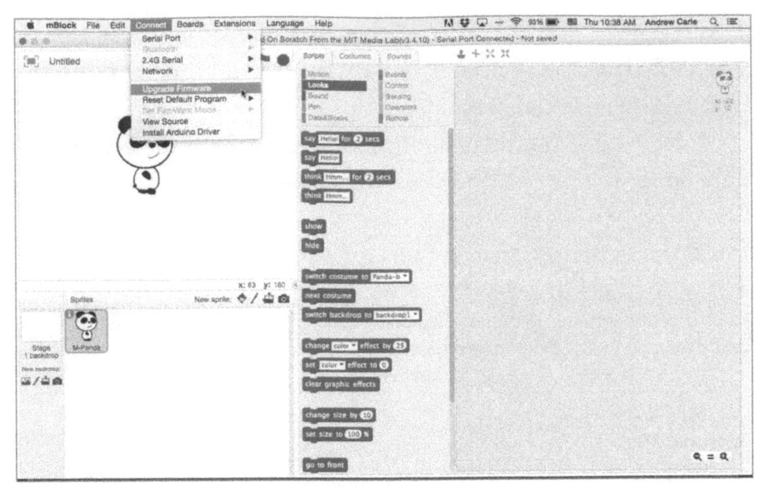

FIGURE 1-13: Update Firmware

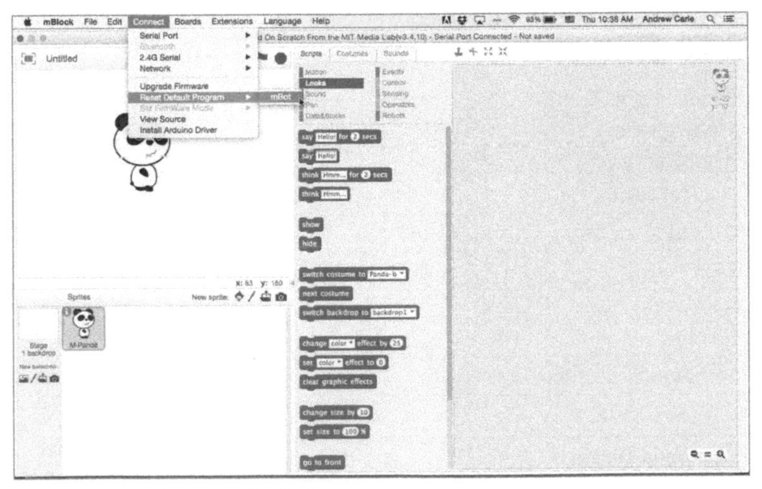

FIGURE 1-14: Reset Default Program

In mBlock, the Reset Default Program option will upload a version of this firmware that includes a *line-following program*, which is a program to avoid obstacles using the distance sensor. It also responds to the infrared remote and the onboard buttons. To fit those extra commands into the mCore's program memory, it trims out support for more advanced Makeblock sensors. This means that if you want to program a robot that uses the Compass sensor or long LED strips, you'll need to replace the Default Program with the Update Firmware command.

Table 1-1 lays out the main differences between these two software options.

TABLE 1-1: Default Program versus Firmware

FEATURE	MBOT DEFAULT PROGRAM	MBOT FIRMWARE
Sound on boot	Three tones	Single chirp
Wired USB connection	X	X
2.4G serial connection	X	X
Bluetooth connection	X	X
IR remote	X	
Stand-alone obstacle avoidance	X	
Stand-alone line-following	X	
RGB LED strips	15 lights	Unlimited
LED matrix	X	X
Seven-segment display	X	X
Temperature sensor	X	X
Joystick input		X
Compass sensor		X
Three-axis gyro sensor		X
Me Flame sensor		X
Me Touch sensor		X
Humidity sensor		X

For more detail, you can read about both programs in the mBlock directory. The IR-supporting version, called Default Program, is in the file `mbot_factory_firmware.ino` and the advanced sensory supporting version is `mbot_firmware.ino`.

WHERE WE'RE HEADING FROM HERE

When you're shopping for electronics kits, it's easy to focus on the hardware specs or potential projects to the exclusion of all else. The mBot has a great set of features that compare well with any other kid-friendly robotics or Arduino system. But the features that bring a smile to my face while working in the Makerspace aren't listed at the top of tech sheets. I love the flexible platform and the small sensible decisions that went into the design of the mBot and mCore as physical objects, ready for oodles of kid abuse with a minimum of adult intervention.

In later chapters, we'll see those same design principles appear when we dive deeper into LEGO integration, mixed media puppets, and large- and small-scale projects. But all of those projects rely on having programming tools that make the powerful hardware accessible to kids of all abilities. Chapter 2 will dive deep into the software for both computer and tablet to demonstrate the power of the mCore. Also, we'll survey the many external sensors that can be connected to the mCore that will be used in the projects throughout the book. From Chapter 3, "Animatronics," onward, we'll combine programming in Scratch with the joy of using sensors to create everything from whimsical creatures that react to their environment to remote untethered data-logging devices to a ping pong ball–flinging robot, ready for battle.

mBot Software and Sensors

The mBot is built atop several well-established open platforms, and benefits from decades of development. While this pedigree means the mBot is fantastically capable, simple questions like, "How do I control my mBot?" can have frustratingly long answers filled with branching paths and "Yes, but . . ." answers. In this chapter we'll cover the entire range of control options for the mBot, from the supplied infrared (IR) remote, to wireless control from a computer or tablet. We'll end with fully uploaded, autonomous operation.

DEFAULT PROGRAM OPTIONS

The mBot arrives out of the box programmed with three different modes, controlled by the IR remote. You can tell this program is currently loaded because of the distinctive three beeps when you flip on the mCore.

Using the factory-installed program, you can steer the mBot with the IR remote's arrow keys and adjust its speeds with the keypad. Pressing the A, B, and C letter keys will shift the mBot between several behaviors. These distinct navigation modes make use of the Ultrasonic Distance sensor and Line-Following sensor that are part of the standard mBot build. The mBot defaults to Mode A, which is the simple steering system just described. Pressing B shifts the mBot into obstacle-avoidance mode, which uses the distance sensor. Pressing

C moves the mBot into line-following mode, which makes the mBot look for and follow a black line underneath it. The mBot retail kit includes a simple paper oval, but the sensors will recognize courses made from dark-colored masking tapee or electrical tape. You can return to manual driving mode by pressing A. The mBot cannot be in more than one mode at a time; for example, there's no way to have the mBot follow a line and avoid obstacles at the same time.

IR remotes are cheap and have many drawbacks, quite a few of which affect the mBot. Infrared requires a line of sight between the remote and the receiver mounted on the mBot. This makes robots heading down the hallway away from the driver difficult to control. Anyone in an environment with many mBots will discover that any of them will respond to commands from any remote.

FIGURE 2-1: It's great that the mBot offers IR remote control out of the box, but it can be a frustrating experience.

In fact, this is one reason why we wrote this book. We've met too many people who confused the limitations of the default program and IR remote (see Figure 2-1) with the capabilities of the entire platform. Building original creations with the mBot requires moving beyond the IR remote to either a computer or mobile device.

MAKEBLOCK APP

Makeblock has improved the quality of their mobile offerings over the last two years, but not always in the cleanest fashion. Apple and Android app stores each have many outdated programs listed, and most have very similar names. At the time of writing, the only mobile app under active development for both platforms is the Makeblock app.

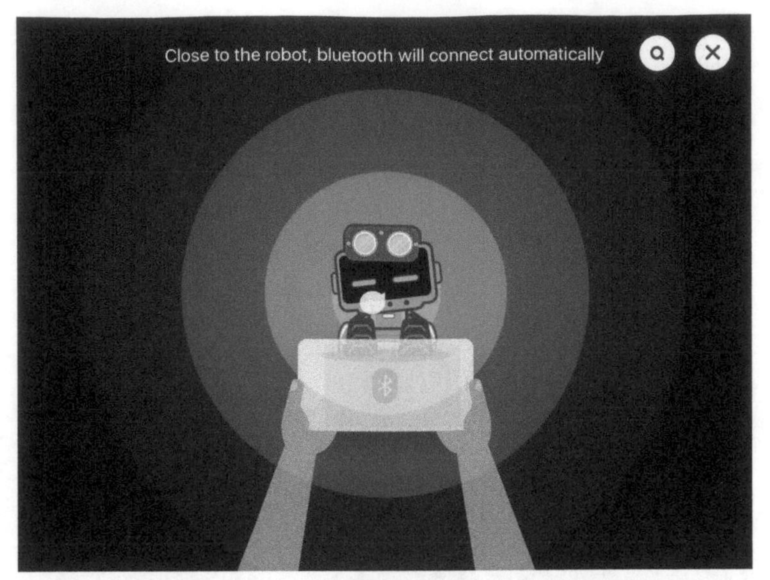

FIGURE 2-2: If there's a single robot turned on and close to the mobile device, Bluetooth pairing can happen in the background. Otherwise, just touch the robot with your mobile device.

Makeblock (the app) supports several robot products from Makeblock (the company) beyond the mBot. When you launch Makeblock, it automatically tries to pair your mobile device with the closest Bluetooth robot. If several robots are in range, the app will ask you to move closer to your chosen robot.

TOUR OF THE PROJECT GALLERY

Once the robot and app are paired, the app reveals a gallery of robot configurations. Each icon contains a customized control interface for the mBot or other robots from Makeblock's product line.

The Project Gallery (see Figure 2-3) shows a line of Official Projects, each based on a particular mBot configuration. The Playground and mBot projects (also shown in Figure 2-3) need only the materials provided in the retail mBot kit. Other projects, like the Cat Searchlight and 6-Legged Robot, ask for extra sensors, servos, or metal Makeblock parts. These extra requirements display an orange Expand label on the top right of each project icon. You can view required materials and build instructions for Official Projects by clicking the info icon on the top left of the Play screen (shown in Figure 2-4).

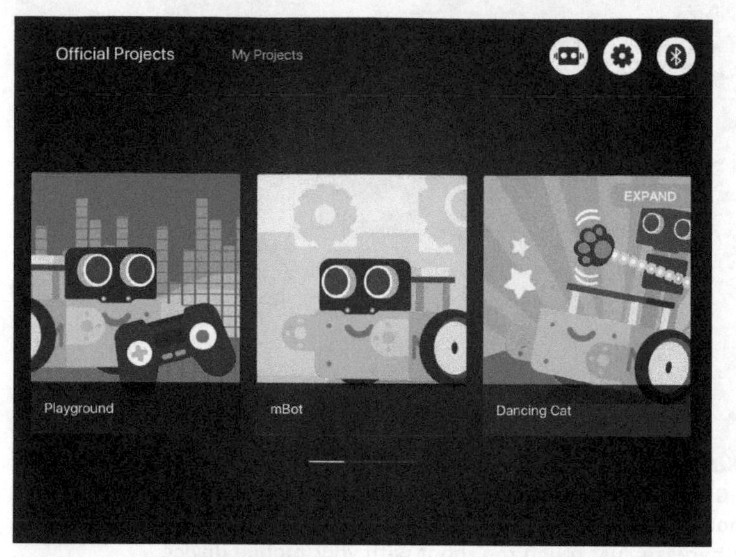

FIGURE 2-3: The Makeblock app's Official Projects expect robots built exactly as specified in the linked instructions. Changing any element will move the project into the My Projects section.

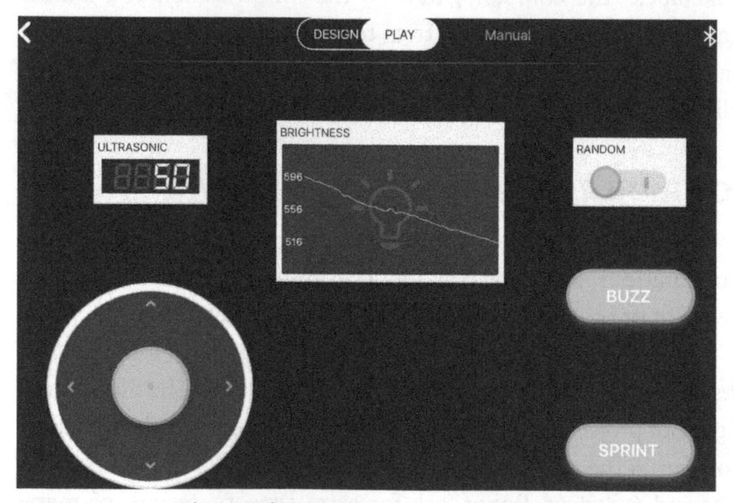

FIGURE 2-4: In Play mode, some screen elements control the connected robot, like the D-Pad or the Buzz button. Others display live data from the mBot sensors.

Touching any picture in the gallery opens a control panel built for that configuration. That includes sensor displays, buttons to trigger specific behaviors, and control tools for motors or servos.

This is like the LEGOs Robot Commander app, which offers the same sort of drag-and-drop control schemes for different LEGO builds. However, when users move from Play to the Design tab (see Figure 2-5), the Makeblock app offers far more control over the tools. (See Figure 2-6.)

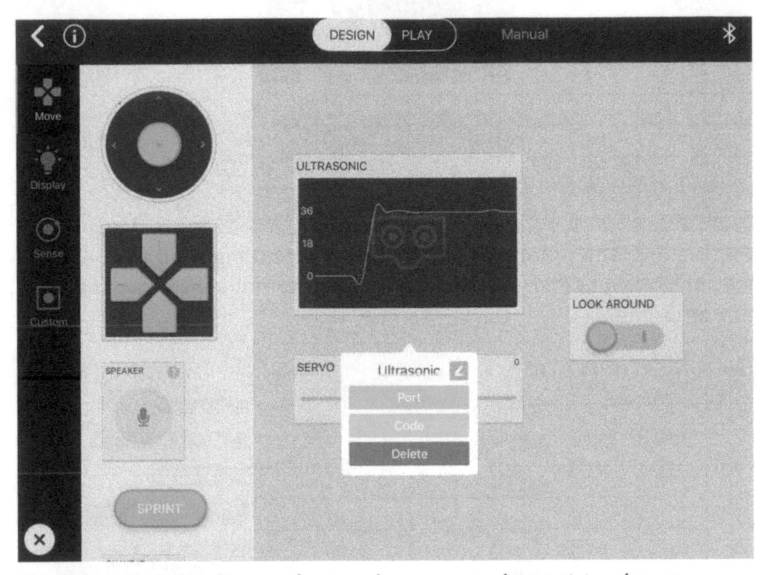

FIGURE 2-5: In Design mode, touch a screen element to change which port a sensor connects to or modify the code for that widget.

To get even more control than the Design tab, you can open the code attached to the screen widget and make more fundamental changes. Each control element is a front-end, block-based piece of code based on Google's Blockly libraries. The gallery on the left edge of Figure 2-7 contains all the blocks necessary to change any controls or displays currently on the screen or create new ones.

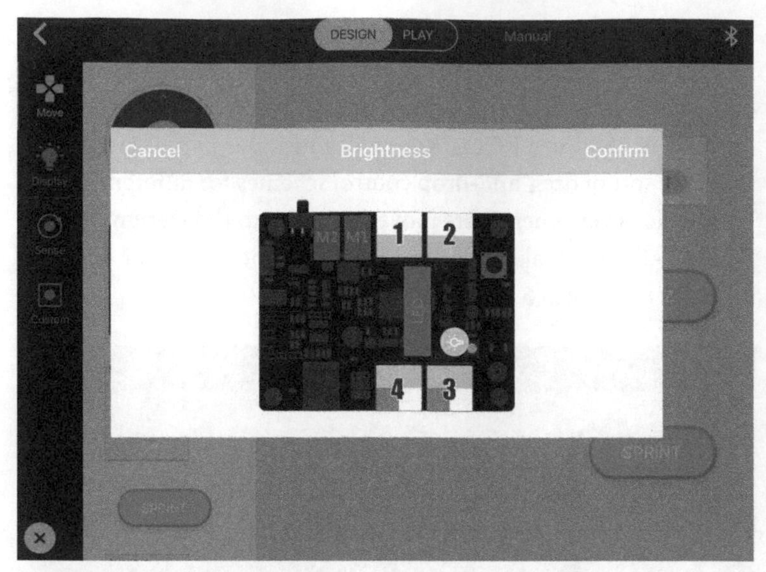

FIGURE 2-6: The mBot can get a brightness reading from the onboard sensor, shown as the yellow circle, or an external sensor connected to port 3 or 4.

FIGURE 2-7: Blocks in Makeblock app's Begin, Move, and Display palettes

NAVIGATING BLOCKS ON A MOBILE DEVICE

The selection of blocks in the Begin palette changes for each type of UI element. Single buttons only have "when pressed" and "when released" options. The D-Pad controller has a "when pressed" and "when released" option for each of the four directions. Numeric displays and graphs only offer a "when start" option.

Direction blocks in the Move palette assume the standard mBot motor configuration. They also allow direct control over individual motors or servos.

Purple Display blocks (see those shown in Figure 2-7) allow control over physical LEDs, sounds from the mBot's speaker, or elements on the Makeblock app screen.

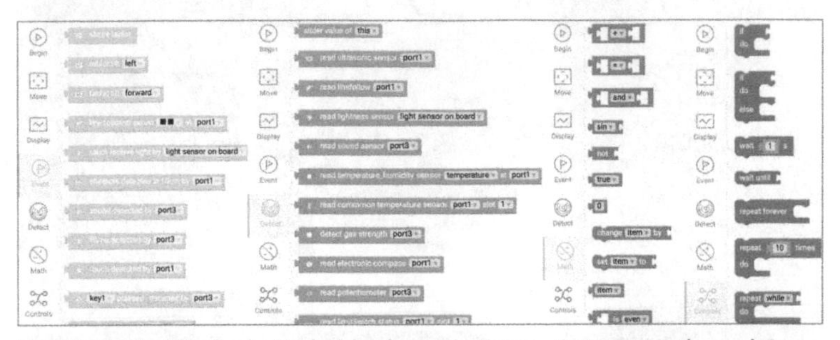

FIGURE 2-8: Blocks in Makeblock app's Event, Detect, Math, and Control palettes

Event blocks look for input from attached sensors or the mobile device. Using these blocks, it's possible to create a simple system that steers the mBot around by tilting the mobile device. This is a great opening challenge, but in our experience, kids quickly determine that the latency between the Makeblock app and the robot makes for a frustrating drive. The Detect palette provides specific blocks for most sensors sold by Makeblock. All of these are puzzle-piece shaped blocks, which means they connect with other blocks in the program and provide the numeric value of the given sensor.

Math blocks bundle all the essential arithmetic operators and functions. They also control the Makeblock app's implementation of variables. We explore these blocks in some detail in Chapter 4, "Measurement Devices."

Finally, the Controls palette holds all conditional statements and Wait and Repeat loops.

We started with the pre-configured mBot control program from the gallery. When we change any screen element in that program, the Makeblock app automatically saves it and asks to rename the project. Anyone can fiddle around with the pre-built robots in the gallery with full confidence that they won't destroy the templates.

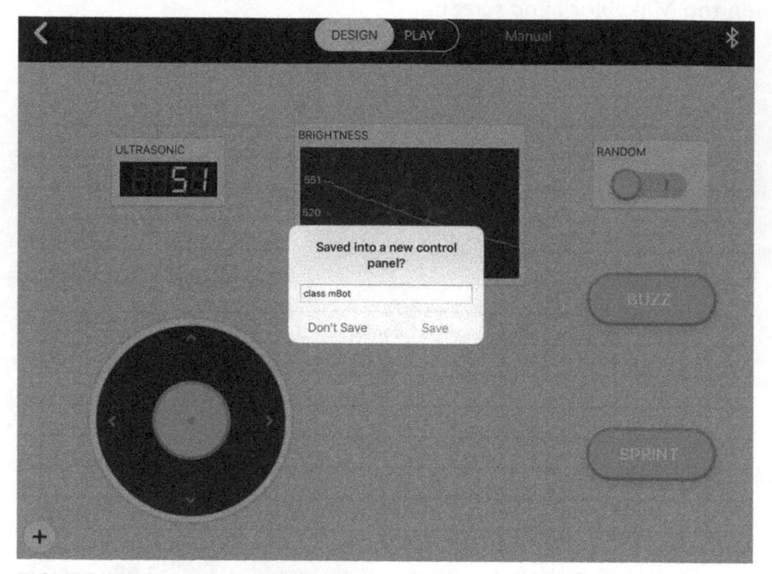

FIGURE 2-9: Once renamed and saved, these modified projects will appear in the My Projects gallery.

Recent versions of the Makeblock app added the Playground project to the Official Projects gallery. Playground is a slick showcase for the mBot's different possibilities, but you can't expand or build on what's provided.

The Game Controller screen (shown on the next page) provides an analog joystick for precision mBot steering. It also has video game–inspired buttons to make the mBot sprint, spin, and shake.

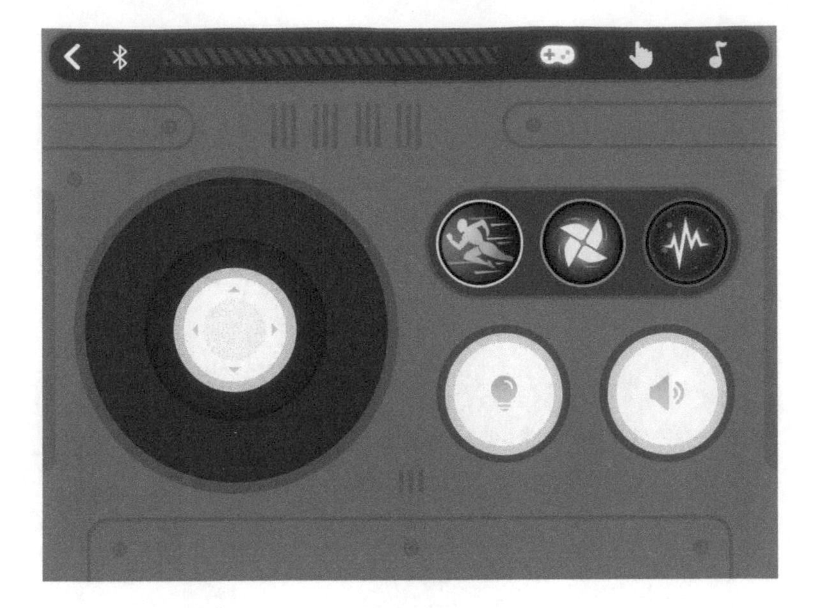

Exploring the Music panel will reveal the limits of the mCore speaker. Clicking the finger icon shown at the top right of the panel activates the Draw-a-Path tool (shown on the next page), which allows even young children to create an independently moving mBot. If you draw a path in the box and hit the Play button, the mBot will dash off and follow that course! You will see the mBot's progress along the path shown on the screen. The active zone is about a 1 m × 2 m rectangle. Since tables, chairs, and other real-world obstacles don't appear on the Draw screen, collisions are pretty common. Even so, the Playground Draw-a-Path tool is a fun new option in the Makeblock app. It's found a great home as part of Balloon Tag!

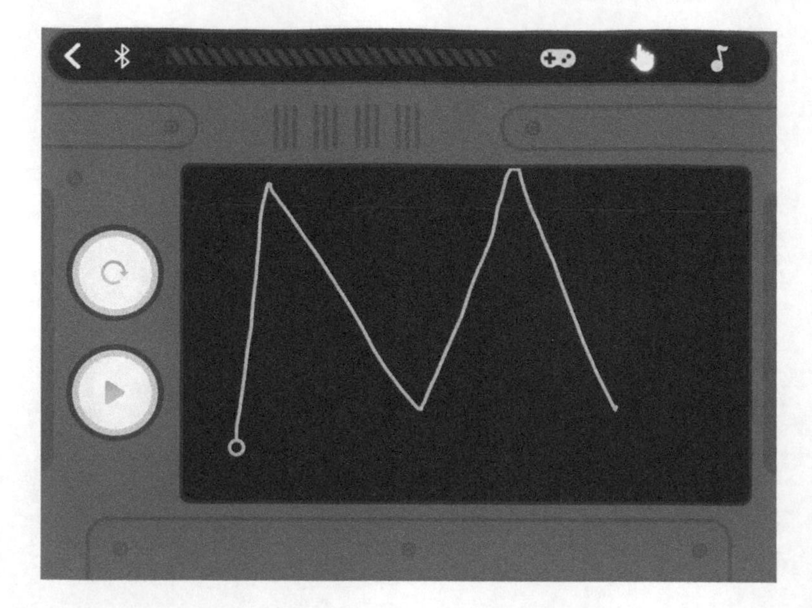

BALLOON TAG

Using mobile apps opens up a world of multi-mBot games and activities that are impossible with the IR remote. One of our favorites is the mBot Balloon Tag. This is a flexible activity that's anchored by the sheer chaotic joy of popping someone's balloon. The materials list is self-evident: you need a balloon for each mBot and a sharp thing with which to pop the balloon. You can establish a super-serious league for this game, with standardized bots and balloons to better focus on pilot skill—but that's not how it works in our classrooms. Our focus is more on the design and engineering aspects of the challenge.

Prep

Provide each group with an mBot, several balloons, and a lance of some sort. We've had success with wooden BBQ skewers, plastic straws with thumbtacks, or even sharpened pencils.

Depending on the age of the students, it's a good idea to specify where and how the balloon should be attached to the mBot. In its base form, the mBot lacks good mounting points parallel to the ground or along its central axis. One way to create these points is by adding two right-angle Makeblock brackets and some Makeblock or LEGO beams to the rear spurs of the mBot frame.

These create a stable, rigid frame that can support much larger structures. Just don't overload the mBot! For light-duty work like a balloon mount, cable ties can work just as well. You can either knot the balloons around the cable tie or connect them with a loop of string. Such a wobbly connection makes the inflated balloon a shifting target in the game.

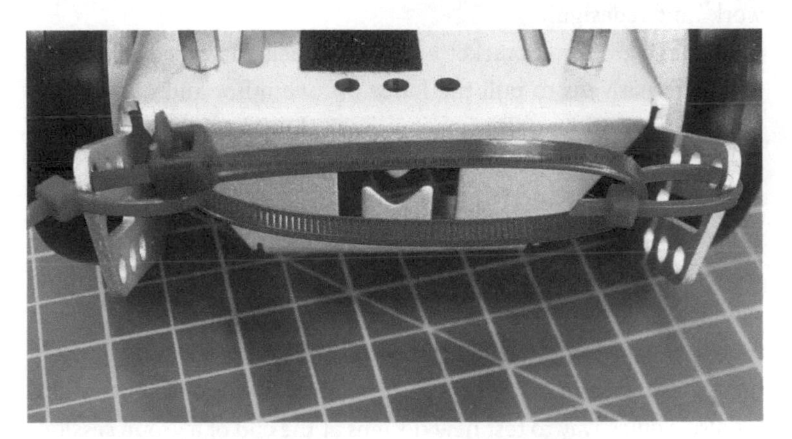

IMAGE COURTESY OF @MISTERHAY

Mounting a rigid lance to the mBot is much more difficult. You can lash it to the brass standoffs with cable ties, or to the frame alongside the battery. This process is full of interesting challenges, most of which aren't obvious to students at first. What angle will allow the lance to best reach the balloons? Will the lance shift from side to side? Will it extend far enough in front of the mBot to push and trap the opponent's balloon? Investigating these questions will lead students to consider outlandish designs. Many of these designs will not work because of the restrictions and requirements of the materials, and the students will have to start again. This is the heart of a powerful iterative design process. Finding answers to these questions is the core of the activity.

It's a good idea to set a short time limit on each joust—giant melees are too chaotic. After each game, devote some time to "pitwork" and redesign.

In large groups, a bracket of balloon duels can take too long. A great alternative is to pull the lance off one mBot and make it the target, with the other mBots becoming the hunters. Allow the person controlling the target to use the Draw-a-Path tool in the Makeblock app Playground project. Hunters and prey will take turns moving. The extra mobility of the Draw-a-Path tool allows the player controlling the target bot to juke around clustered hunters. This can cause some significant pileups. This asymmetric version of Balloon Tag, where each side is using a different control method, works great where there's a short time frame or a fixed number of turns for both sides. It's also a quick way to test new designs at the end of a group session.

If you have access to materials beyond the retail mBot kit, there are even more possibilities. With a servo motor and some clever mounting, you can control either the lance or balloon in the Makeblock app control panel while driving. This drastically increases the challenge level of building and steering. Classroom tests show significantly higher self-popped balloons when servos are used.

Utilizing the default mBot build that includes a Line Follower sensor under the chassis creates some interesting racing variants of Balloon Tag. Instead of a grand robot melee, create a small course of line-following paths with open spaces in between. Teams start with the Line-Following project, but modify it to start with a button press and add the driving controls of their choice. Robots jostle and fight in the open spaces, but need to locate the line and use the line-following mode to travel to the next waypoint. This structure breaks up the mad scrum of normal Balloon Tag with high-emotion chase segments, as the lead mBots rush to the end of each path with their balloons exposed to the crowd.

We call this Balloon Tag to specifically connect to the free-wheeling dynamic games our kids play at recess. See what new ideas emerge from adding a new part, or how altering a rule changes how people play. In each case, the new tools will create more complications, more challenges, and more powerful, student-driven learning.

IMAGE COURTESY OF @ROBOTICS_FUN

MBLOCK

Makeblock's mBlock is a visual programming environment for Windows, Mac, Linux, and Chromebook computers—if you're working on a device that has a screen and physical keyboard, then there's a version of mBlock for you. It expands on all the capabilities offered in the tablet programming apps and provides the most robust tool for programming the mBot.

The mBlock platform is a direct fork of Scratch from MIT Media Lab's Lifelong Kindergarten (LLK) Group, and it inherits Scratch's incredible feature set. It presents robotics commands in a format familiar to millions of young people.

Mitchel Resnick, head of the LLK Group, often describes Scratch as having "low floors, high ceilings, and wide walls." In the programming world, *low floors* means that everyone can enter, with no background or prerequisites.

High ceilings allow users to grow and expand their skills for years or decades before hitting something that "just can't be done." *Wide walls* implies that the tools should allow as many different types of creative expression as possible. Scratch does that—it provides the tools to make everything from anime music videos to multiplayer platformers.

High school students often scoff at Scratch and other block-based languages as "programming for kids." This reflects their own inexperience rather than the potential of block-based programming. The mind-blowing projects from Scratch user "griffpatch," or coming out of UC Berkeley's Beauty and Joy of Computing course, should shatter that illusion. This is not the last time you'll hear us say, "Simple doesn't mean easy."

The mBlock platform is a natural extension of the house Scratch built. It adds an extra room for physical robotics, without disrupting the existing floor or ceiling.

If you're interested in the non-robotics potential of Scratch, there is a great library of books waiting for you. We recommend:

» *Make: Tech DIY: Easy Electronics Projects for Parents and Kids,* by Jaymes Dec and Ji Sun Lee (Maker Media, 2016)

» *The Invent to Learn Guide to Fun,* by Josh Burker (Constructing Modern Knowledge Press, 2016)

» *The Big Book of Makerspace Projects: Inspiring Makers to Experiment, Create, and Learn,* by Colleen Graves and Aaron Graves (McGraw-Hill Education TAB, 2016)

» *Coding Games in Scratch,* by Jon Woodcock (DK Children, 2015)

Although this chapter will teach you how to build functioning programs from a blank screen, there are many discrete worlds to explore in Scratch. Dive deep in some other areas and see how much that exploration adds to your robots!

CONNECTING TO MBLOCK

As of publication, the current version (v3.4.11) of mBlock for Windows, Mac, and Linux computers bundles the Scratch-based block environment and the Arduino tools into a single platform-native program. There is also a web-based tool, available at *http://editor.makeblock.com/ide.html*, which provides the same toolset within a modern browser. Beta versions of mBlock 4.0 suggest that, going forward, Makeblock will abandon the different versions for Windows, Mac, and Linux in favor of a downloadable version of the browser-based tool. Since the functionality with each of these versions is nearly identical, all of the programs or projects in this book should work on any future version of mBlock. However, the operating system–specific instructions for connection may change over time.

Every time you open mBlock, you'll need to connect the board to the software using one of three possible connections: Bluetooth, 2.4G wireless serial, or USB. All retail mBlock kits have USB ports and one wireless connection. If you bought mCore boards without buying the mBot kit, you'll only have access to USB. The wireless modules are for sale from Makeblock, and they're easy to swap between boards. If you're using both serial and Bluetooth connections, remember that you identify the Bluetooth boards from a distance by the copper antenna shown in Figure 2-10.

A Word about Connection Types

On a small scale, there's not a huge difference between the Bluetooth and wireless serial connection. If you're considering a larger scale mBot army where you'll work primarily or exclusively with laptops, we strongly recommend the 2.4G serial adapters. In the worst-case scenario, when kids have ignored our color-coded stickers and mixed up mBots and the paired USB dongles, the 2.4G

FIGURE 2-10: In an environment with both wireless serial and Bluetooth hardware, the printed squiggle antenna on the Bluetooth board helps distinguish between the two tiny boards.

serial module has a super-clear indicator when this unit's best beloved dongle is plugged in nearby. This means that I can troubleshoot most connection problems from across the room, without ever seeing the laptop's screen.

Bluetooth may offer maximum flexibility for a single mBot unit, but 2.4G serial is the best choice in any environment where students will work with multiple mBots and computers.

Connecting Bluetooth for Windows

Connecting with Windows is easy. Make sure the Bluetooth module is installed on your mCore, turn on the mCore, and launch mBlock. Make sure your computer has Bluetooth enabled. Click on the Connect menu, then Bluetooth, then Discover.

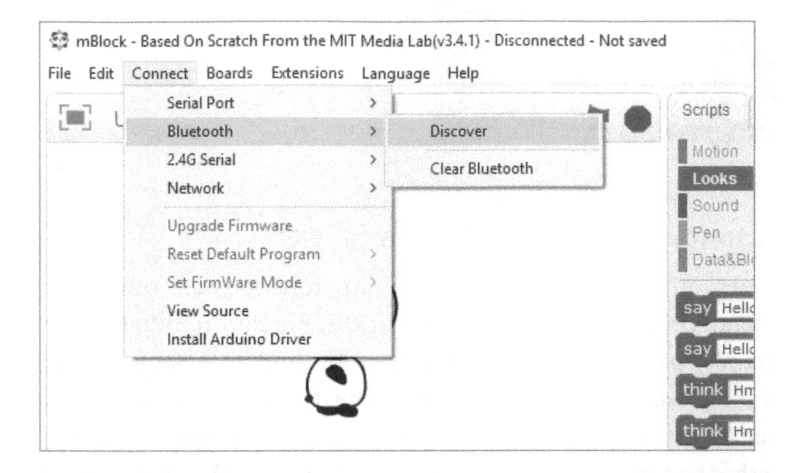

When your computer discovers your Makeblock Bluetooth module, the following screen will pop up with the specific address of that Bluetooth module.

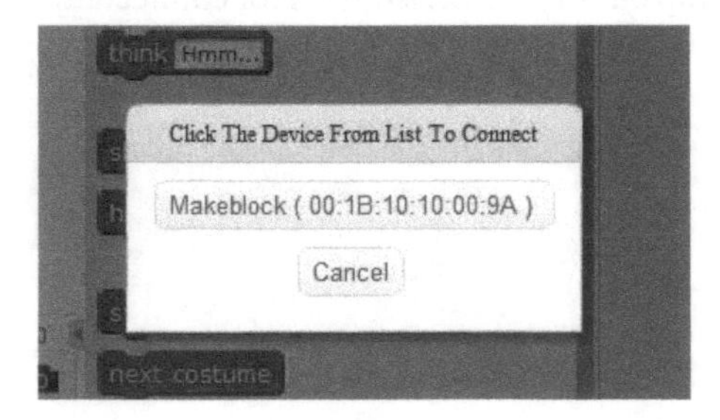

Click that device and you'll get the confirmation message shown in the following image. You are now connected and ready to begin programming!

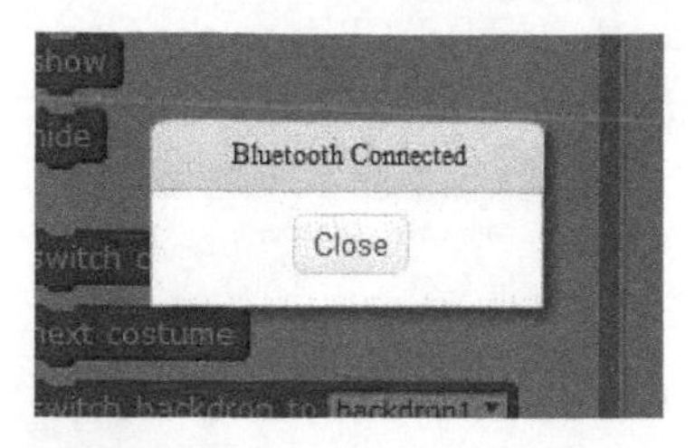

Connecting Bluetooth for macOS

Bluetooth devices need to be paired before software can access them. On macOS, that happens in the Bluetooth System Preferences panel. Make sure the Bluetooth module is connected to the mCore, and then turn it on. After three beeps, you should see a new entry show up in the Bluetooth control panel. This can either be a messy MAC address or a well-named Makeblock entry. The non-human-readable name should only appear the first time you connect to a new Bluetooth module. Click PAIR on the new device.

That entry will quickly flip back to a worrisome "Not connected." Don't stress!

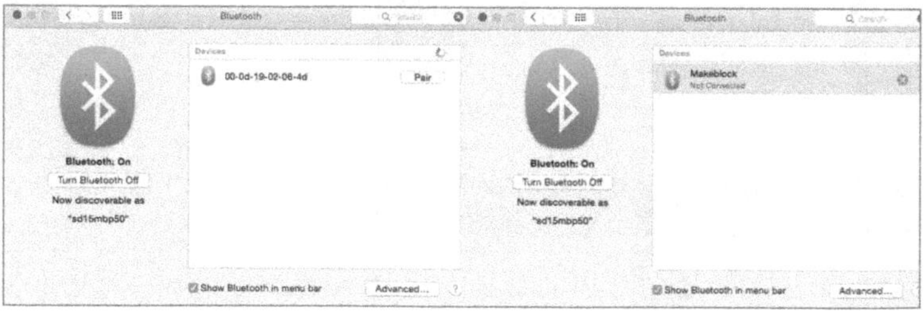

FIGURE 2-11: When pairing a Bluetooth module for the first time, the MAC address appears instead of the Makeblock name.

With this accomplished, return to mBlock and open the Connect menu. Despite the existence of a Bluetooth item in that list, you need to open the Serial submenu and then choose the new `tty.Makeblock` entry. (Yes, this is a mess.) With a top-level Bluetooth menu that stays grayed out and two permanent entries in the Serial menu that use the word Bluetooth, the important thing to click is `/dev/tty.Makeblock-ELETSPP`.

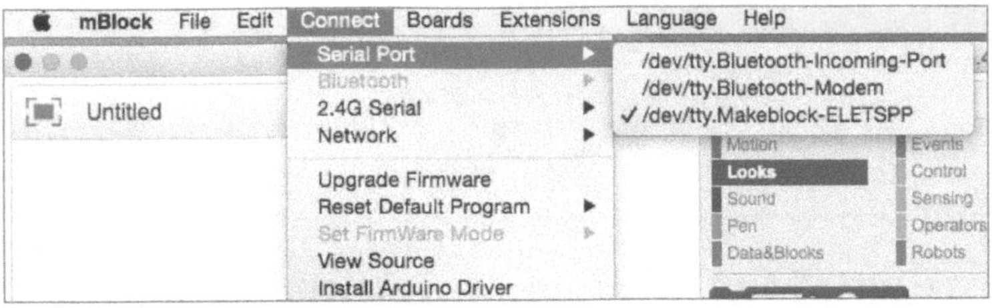

When that works, you'll see a small check mark appear by the `/dev/tty.Makeblock-ELETSPP` entry and the window header will show Serial Port Connected.

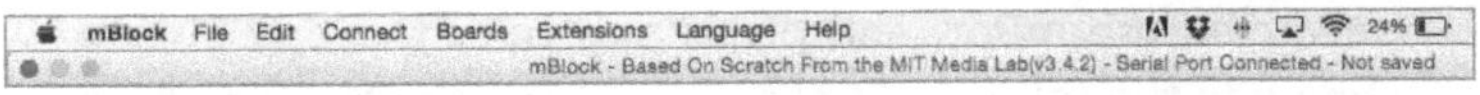

Connecting 2.4G Wireless Serial

The crucial thing to remember when using the wireless serial connector is that each USB dongle and the small communications board that ship together are paired *to each other*. Don't throw all of the USB dongles in a drawer! Use a bit of Velcro to attach the dongle to the mBot frame when not in use. If you're working in a classroom setting, pull out the sharpies and stickers and label them posthaste!

If you power on the mBot when the USB dongle is not attached to a nearby computer, a tiny blue LED on the communication board blinks. This blue light will glow steady within seconds when the dongle is connected. Keep this in mind if you ever have to sort through a large pile of mismatched components.

Once the mBot is powered and the USB dongle plugged in, it just takes a single click to connect the board to mBlock.

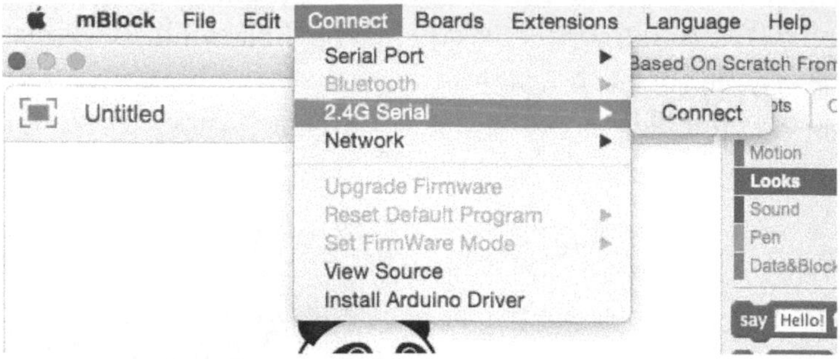

Paired for Life?

Although, in the classroom, we insist that the USB dongles and serial boards are paired for life, that's an exaggeration. There's a button on the serial board that will forcibly pair it with a USB dongle in range. If you're somehow stuck with a mismatched set, plug the USB dongle into a computer, power on the mBot, and press and hold the tiny button shown in the following image. You'll see the flashing blue LED turn glow steadily after a few seconds. However, like matching socks in the laundry, when you make a new match you're also creating two other broken pairs.

When the connection is active, the status message in the top bar will change.

mBlock - Based On Scratch From the MIT Media Lab(v3.4.2) - 2.4G Serial Connected - Not saved

Connecting USB

Although USB is an incredibly familiar technology, there are two points worth noting about the mCore's USB connection.

FIGURE 2-12: The mCore uses a USB-B plug, the style often used for printers. It's sturdy and can take a beating.

First, the power switch on the board needs to be on in order connect to mBlock over USB, whether a battery is attached or not. This goes against safe practices for normal Arduino boards, which can receive power from either an external source or USB, but not both at once. The mCore board's design prevents this "two power source" problem. If there is a rechargeable battery attached, plugging in the USB cable while the power switch is off will charge the lithium battery.

Second, the mCore board uses a USB-to-serial chip that's common to a Chinese-made Arduino clone known as the CH340. This chip requires the installation of a specific driver. If you connect the USB cable and don't see a new entry appear in the Connect → Serial Port menu, check to see if you are missing this driver.

The Install Arduino Driver item in the Connect menu will install the CH340/CH341 driver for your platform from within mBlock.

Note that this requires admin access on most computers, so it can be tricky to do with student machines. This is only required when using a wired USB connection to the mCore.

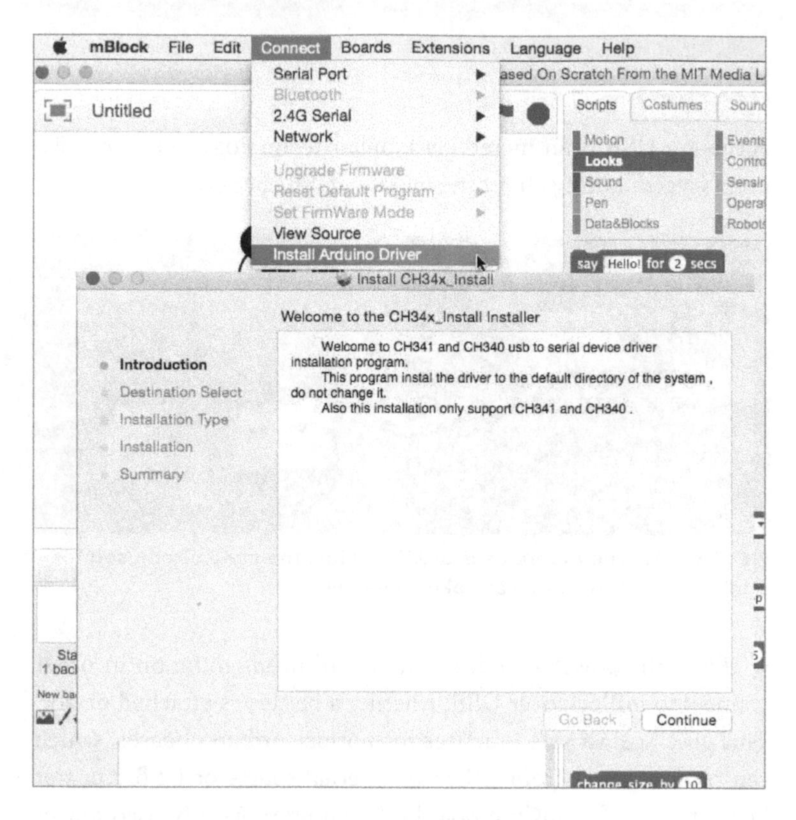

Keeping the previous two notes in mind, opening a USB connection is simple. Connect the board to the laptop, make sure the power switch is set to ON, and select the proper serial port from the menu. On Windows machines, this will be COMx; on Macs it will be in the form `/dev/wchusbserialXXXX`.

So far, we're using all these as tethered connections, even though two are wireless. *Tethered* simply means that the program logic stays on the computer and is sent to the mCore board over this active

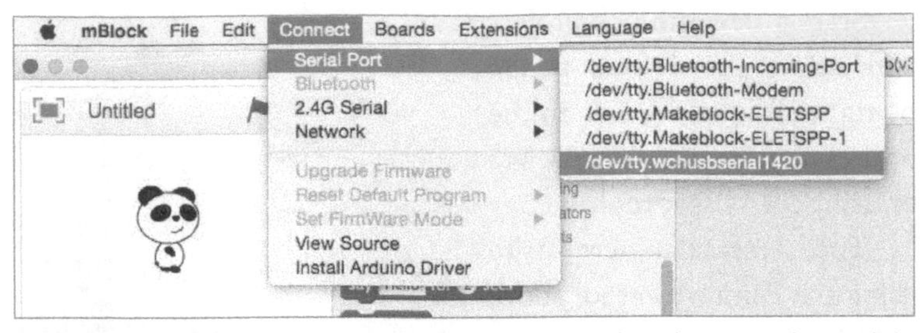

FIGURE 2-13: The wired USB connection is the last item on the serial port list.

connection. There's a constant two-way stream of instructions and sensor data between the computer and robot. If this communication is disrupted while the mBot is battery powered, the robot will continue to perform the last chunk of the program sent by mBlock. When this disruption happens because the mBot moves out of range of the Bluetooth or 2.4G signal, this can cause strange behaviors that don't scream "out of range." Restore the connection by connecting the hardware or bringing the mBot back within range, and restart the mBlock program.

TRAFFIC LIGHT CLASSROOM VOLUME METER

Traffic light volume meters are a staple of teacher supply catalogs. The various LED units make this a very accessible physical project.

There's a real value to prototyping physical systems using the sprites in mBlock. Digital prototyping separates the programming logic from construction and wiring, and allows students to focus on the behavior. The version presented here uses the computer's microphone to measure volume at first, instead of immediately bringing in the mBlock's Me Sound sensor. In our classrooms, first-draft prototypes normally rely on sensors on the computer, or even Scratch variables that represent ideal sensor data, instead of mBot hardware.

Start a new mBlock project and delete the default panda by right-clicking the icon in the Sprites panel and clicking Delete or using the scissor tool.

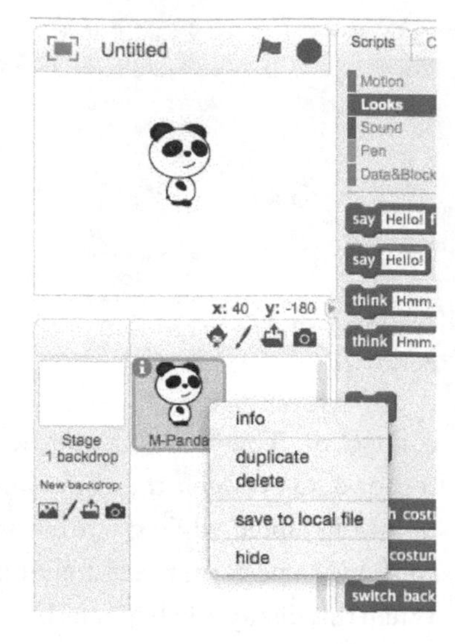

Then, create a new sprite using the Paintbrush tool above the Sprites panel. Change to Vector Mode in the image editor (this will move the drawing tools to the right edge of the screen) and create a simple, filled, gray rectangle. Vector mode will allow us to easily resize this shape later to fit around the green, yellow, and red traffic lights.

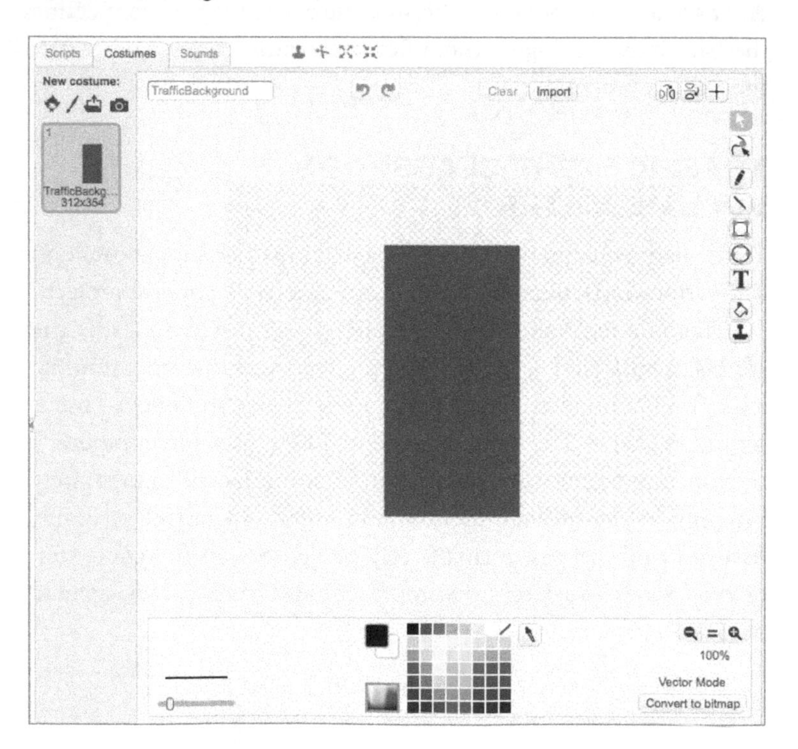

Next, click the blue arrow to access the details for this sprite. Rename the object now, as a way to model best practice for your students. Don't wait until you have a confusing muddle of Sprite 1 through Sprite 16. Do it now. Now.

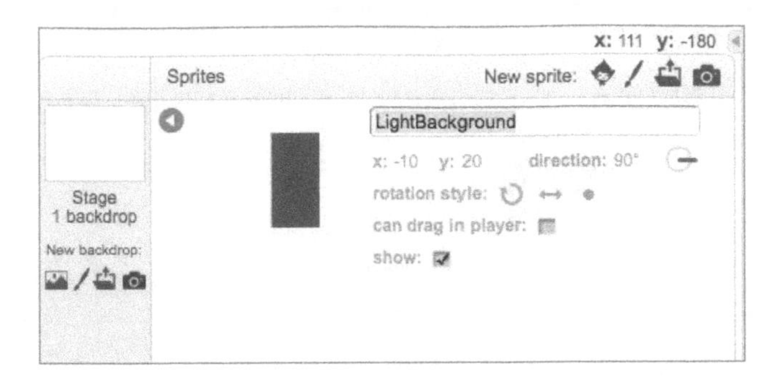

As an offshoot of Scratch, mBlock has a sizable library of sprites and backgrounds, as shown in the next image (although it doesn't look sizable). We'll use one of these as the basis for our three traffic lights. Click on the tiny creature in the New Sprite bar and choose Button 1 from the Things group. This works fine for the green light, but we'll need to copy and recolor it for the other two.

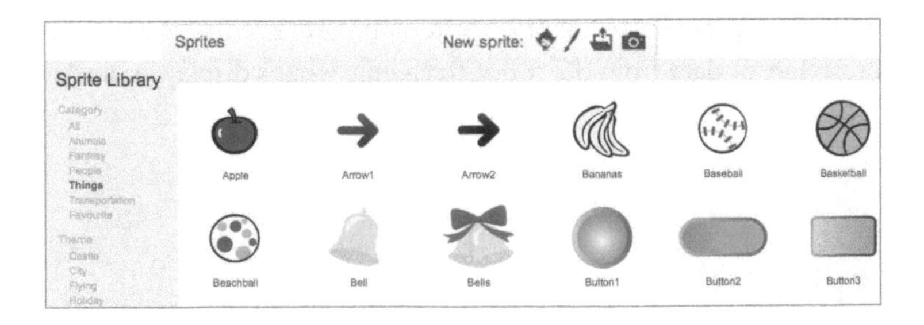

First, duplicate the button twice by right-clicking it on the Stage or in the Sprites panel. Then, rename all three buttons to show the color each one will become.

Select the RedLight in the Sprites panel and then open the Costumes panel. The Button sprite was already a vector graphic, so all we'll need to do is recolor the gradient using the Vector Bucket. Choose two yellowish colors that work for the light and click away.

Notice that the Button sprite has two shapes that each need to be recolored.

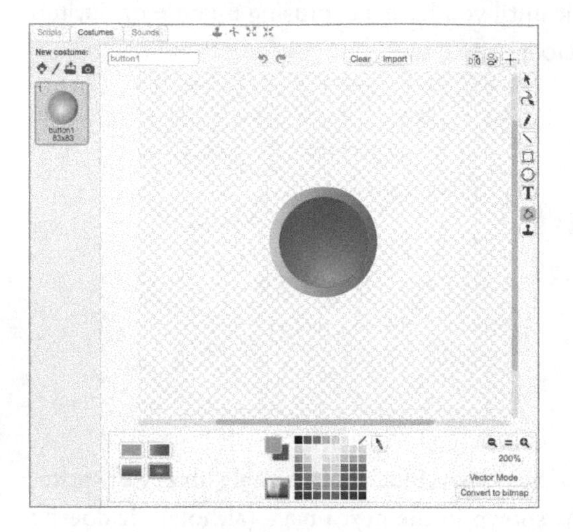

Now it's time to add code to our project. In mBlock, like in Scratch, each Sprite (and the Background!) has a Scripts panel for code that controls its behavior and appearance. When you are writing code designed to control a mobile mBot, it makes sense to keep all of those scripts collected in one Sprite. A project like this is designed to use sensor data from the mBot to change what's displayed on the

```
when clicked
go to x: -10 y: -100
clear graphic effects
forever
    if loudness < 40 then
        set brightness effect to 20
    else
        set brightness effect to -40
```

FIGURE 2-14: This block places the green traffic light on the Stage, and then constantly measures the sound level. The light is bright when the sound is low and dark when the volume rises.

Stage. Therefore it makes sense for each sprite to read data and adjust its appearance. Here's one way that might look for the green light.

The *Green Flag* block, as it's commonly known in Scratch circles, is a basic start-of-program trigger. Every sprite can have its own Green Flag block. In fact, an individual sprite could have several. Having several start blocks allows a sprite to have parallel routines, which can be exceptionally useful. For this first program, however, we'll just use one block.

When the program starts, it's sensible to reset the position and appearance of the sprites. It's not technically necessary here, since these sprites don't move at any point during our program, but it's another good habit to model for students. Like Scratch, mBlock does not have any built-in reset or cleanup. Adding a GoTo block that defines where a sprite should start on the Stage means that that clicking the GreenFlag to restart the program will also undo any accidental clicks that moved the GreenLight Sprite on the Stage the next time the program runs. We need a similar block to reset any changes made to the appearance of the sprites, including size, costumes, or graphic effects like brightness or transparency. Since the program will adjust the brightness of the traffic light sprites to indicate that they're lit, we'll include the ClearGraphicEffects block under the GoTo block to ensure that this light starts dark.

All of the blocks shown in Figure 2-14 execute once, in the order displayed, at the start of the program. Everything that follows is wrapped in a Forever loop, meaning that they will cycle quickly and endlessly.

Next, we will check sound using the Loudness block from the Sensing palette. All of Scratch's sensing capabilities have been passed down to mBlock. Scratch was designed to take advantage of the microphones and webcams built into most computers, including a simple block to measure ambient noise. It's great to make use of these built-in options when starting out with young programmers. By starting out with only software tools, we allow kids to focus on the core ideas of their program before introducing wires and other physical complications. Then, once the ideas are sound, out come the full robots.

An If/Else comparator checks the loudness level against our chosen threshold value of 40. The mBlock Loudness sensor returns values between 0 and 100, so 40 is on the soft side, but not deathly quiet. By measuring and comparing the loudness against a threshold value, we can create different behaviors for the light based on the sound levels.

Instead of making distinct costumes for the lit and unlit versions of each light, we'll use the Brightness control from the Looks palette. Scratch's graphics properties were passed down to mBlock and can be used to modify the appearance of a sprite on the Stage without changing the costume itself. While novel combinations of Warp, Ghost, Pixelate, and the other effect options are key to many great "lose a life" animations, they can also render a sprite invisible and unrecognizable. Use them wisely. All of these blocks can have positive or negative numbers as values, but that won't always translate into an observable change.

In Figure 2-15, the program sets the brightness of the GreenLight sprite based on the reported value of the sound sensor. If the sound level is lower than 40, then the brightness is set to positive 20, or lit. If it's higher than 40, then the classroom is assumed to be too loud, and the green light goes dark, with a brightness of −40.

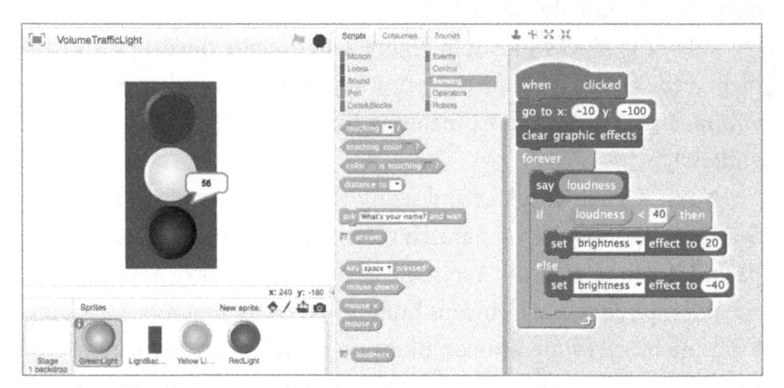

FIGURE 2-15: Say blocks in an mBlock script place a word balloon above their sprite.

Dragging a script block to another sprite will copy that block to the new sprite. This is a great shortcut, but also an easy way to introduce errors. If we copy this script block from the GreenLight to the RedLight sprite, it creates two lights that move in sync, instead of one light that turns on when the room is loud and another that stays lit when it's quiet. Copy the block, but then open the RedLight script panel to make the necessary changes.

The easiest part to adjust is the position of the red light itself. Keeping the X coordinates the same ensures that the lights stay vertically aligned. Of course you can change that design if you're used to horizontal traffic lights.

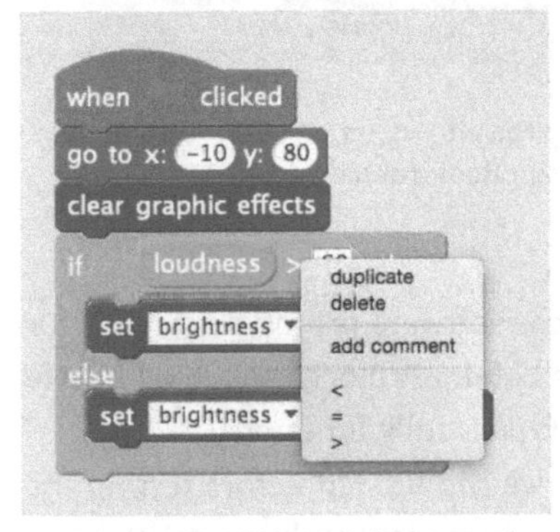

FIGURE 2-16: Right-clicking on the angled green block will allow you to swap between greater than (>), less than (<), or equals.

We also need to change the script so that the sound sensor checks for sound levels above the threshold value. In mBlock, each angled green operator block can change between checking greater than, less than, or equality. When you're revising a program, changing these blocks by right-clicking them can save a lot of time as opposed to dragging new blocks out from the palette.

We can copy and modify the YellowLight script in much the same way. The only wrinkle is that the yellow light needs to be a Goldilocks, only turning on when the sound isn't too soft or too loud. Building logical groups of conditions in Scratch requires <AND> and <OR> blocks. Like the arithmetic blocks, you can stack these really deep. One of the biggest UI hurdles of Scratch variants is that long calculations or conditionals can sometimes stretch beyond the width of the Scripts window. There's a small arrow button on the border between the Stage and the panels that will minimize the Stage and provide some extra width to the Scripts area.

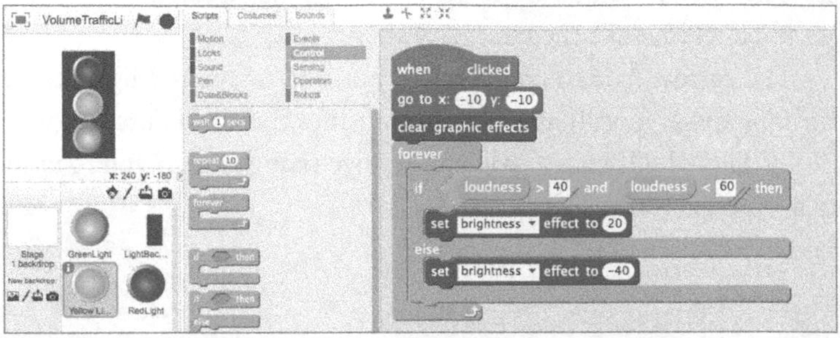

FIGURE 2-17: The green <AND> operator allows us to stack two sensor checks in a single <IF> statement to check for values within a given range.

With that last script in place, the software prototype is ready for testing. Since we've used mBlock's Loudness block instead of the Me Sound Sensor, the software prototype is fully functional.

There's one last helpful block we should add before testing the prototype. Although you can click any sensor block and see the current value, this is cumbersome for something as dynamic as sound. Using a Say block from the Looks palette inside a Forever loop is a great way to stick a sensor value up on the screen. This program has three different Forever loops running, one inside each light, and the Say block works fine in any of them.

So far, this project is fully software-based and uses only components that mBlock inherited from Scratch. Many inspired individuals in our classes are eager to build from day one and grumble over the time spent creating these software prototypes. Fortunately, mBlock and mCore make it easy to grow functioning prototypes into a final physical version, making the next steps feel like incremental revision rather than a blank slate.

Everyone can benefit from working out mistakes and misconceptions in Scratch's low-floor environment where everything *just works.* When working with groups of young people, the software prototype is a critical part of every project. Once a large group starts working with materials, managing those parts consumes a large part of a mentor's or teacher's attention. Once there are cables and batteries strewn over the table, it's difficult to quickly identify whether a problem lies in the hardware or the underlying ideas. Completing a software prototype is a proof-of-concept and provides a touchstone through the rest of the project.

When a project requires a sensor beyond what's included in the mBlock software sandbox, we often connect just the mCore and inputs and model the outputs on the Stage. When particular sensors are in short supply, a prototype can use mBlock variables to simulate the values expected from the sensor. Limited prototypes don't guarantee that the final project will work, but a design that can't work in software is unlikely to thrive with real parts.

Before we push the traffic light classroom volume meter out of the software-only nest, it's worth looking closely at how the whole range of Makeblock sensors interacts with mBlock.

WORKING WITH SENSORS IN MBLOCK

Although there's a long list of different sensors created for the mBot platform when working in mBlock, it helps to think of them as belonging to two basic categories.

Digital sensors measure one thing in the world and report back a binary value: yes or no, on or off, or 1 or 0. Sometimes these are mechanically simple sensors, like a classic push button. In other cases, like the passive infrared motion sensor, the hardware is complex but the value reported back is still binary.

In mBlock, blocks that have a binary value are elongated hexagons. Only blocks of this shape can fit in the question spots of conditional loops.

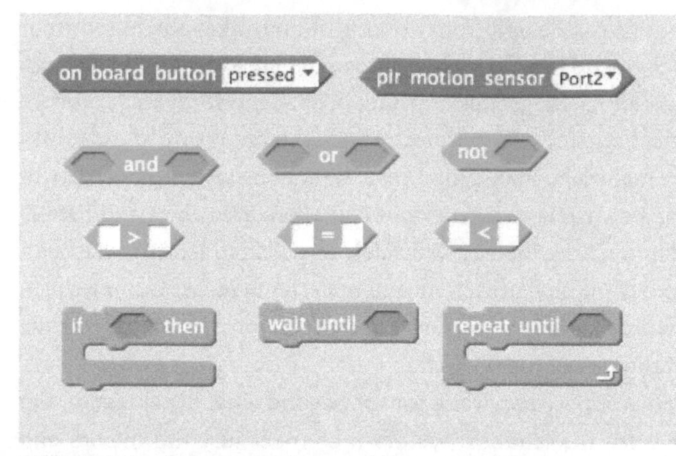

FIGURE 2-18: Green operators or blue sensor blocks with six sides report binary values, just True/False or 0/1.

Despite the shape, these blocks can also be placed into the round openings in Arithmetic blocks and the ever useful Say blocks. This inconsistency is bothersome, but frequently useful.

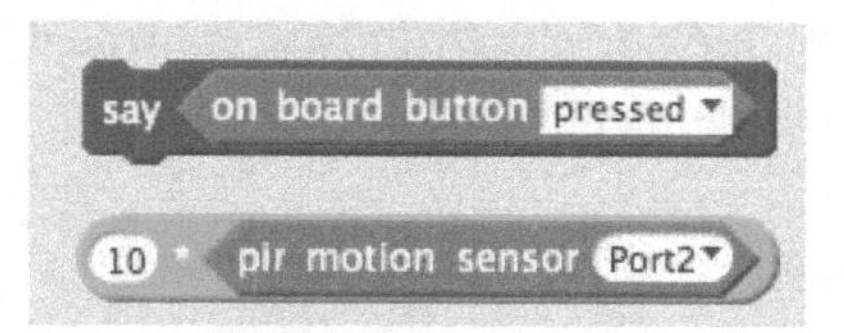

FIGURE 2-19: Using a hexagonal block in an arithmetic operator can create a particular value when the sensor reports true, and 0 when the sensor reports false.

Analog sensors, the other type of sensor, report back their measurement to mBlock as a range of numeric values, normally whole numbers (but not always), and normally positive (but not always).

How these sensors arrive at their values can vary wildly, but at the base level they all are reporting information by adjusting the voltage on the wire. Most microcontrollers, expecting just a binary signal, have a limited ability to read analog signals: the ATmega328 microcontroller, the heart of the Arduino Uno and the mCore, can only read analog signals on ports 3 and 4. Makeblock color-codes analog sensors and ports with dark gray (possibly the least useful color to apply to black plastic parts!).

All analog sensor blocks are ovals, meaning that you can use the values from them in any place you would write a number or place another oval block.

The last pseudo-category of mBlock sensors bundle multiple channels of information together into a single physical package. The most familiar for kids is the Me Joystick, which is a standard analog thumbstick similar to the ones found on every video game controller since the Nintendo 64. In mBlock, the block for the Me Joystick will report on only a single value, either the X-axis value or the Y-axis, at a time. A single block can never report both values, but you can bundle them together into a larger statement as shown in Figure 2-20.

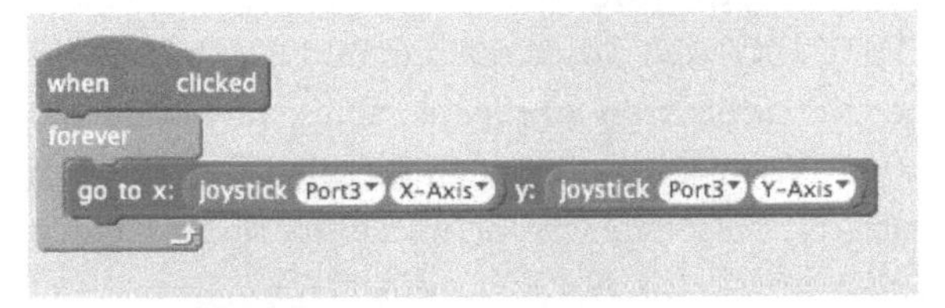

FIGURE 2-20: This single block allows the Me Joystick to move a sprite around the Stage.

Other bundled sensors in the Makeblock line include the mBot's line-follower (two digital light sensors) and the Me 3-axis Accelerometer and Gyro Sensor, which reports three different angle readings.

SENSOR RECIPES

To interact with the real world, an mBlock needs to describe what behavior should occur when the program receives the sensor data. The traffic light volume meter program shows one basic example—each component measures the sound sensor constantly and switches between two states as a result.

This is a common stumbling point for new programmers in Scratch or mBlock. Often, they can describe the behavior they want in very broad terms ("the spaceships, like, shoot all the bugs and they go SQUUUISH!!") but lack the vocabulary and experience to break that complex action into smaller components.

Although students can often describe the behavior they want, it takes familiarity with basic programming concepts and the mBlock environment to see how blocks might combine to create those behaviors. To help bridge that gap, we've included a short recipe list that catalogs basic ways to tie sensor data to outputs, labeled with kid descriptors alongside more technical terms.

Exploring these models should help learners begin to develop an understanding of the way imagined behaviors might look in blocks or code. Recipes are useful tools to help in transitioning from simply observing the sensor value to creating a system that *uses* the data.

Block-based programming reveals the visual structure of these programming concepts—structures that can work with any sensor and any output. For clarity, we've used the onboard light sensor for analog values, the onboard button for digital values, and M1 motor for a generic output. When reading through these recipes, think of the inputs and outputs as placeholders for any sensor or output you want to work with.

"Wait for the Sensor Reading to Hit a Value and Then Do Something," Also Known as *Latching Trigger*

This is the classic intruder alarm from movies. This loop sets a behavior, and then constantly checks the sensor and compares it against a threshold value. Once that threshold is crossed, the behavior changes and *never changes back*.

```
when clicked
set motor M1 speed 255
forever
  if light sensor light sensor on board < 600 then
    forever
      set motor M1 speed 0
```

"Whenever the Sensor Hits a Value, Do Something Until I Say Stop," Also Known as *Latching Trigger with Reset*

While it sometimes feels like the car alarm outside your window will never shut up, most do include some form of a reset button. Building on the previous code block, this script adds the ability to use another sensor to reset to the first behavior.

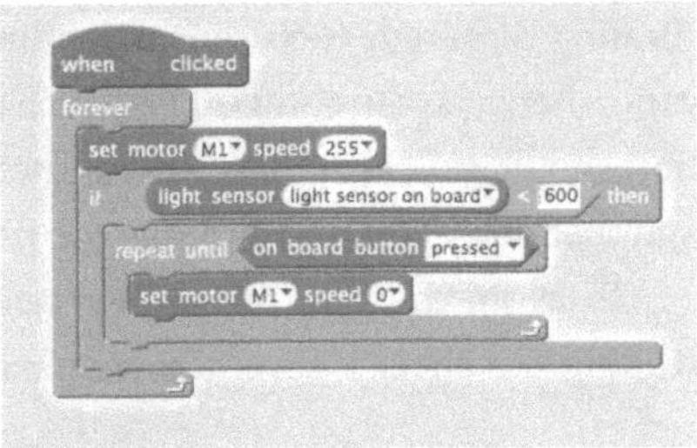

"Constantly, Based on the Value, Do This or That," Also Known as *State Check*

This is the same sort of check used in the traffic light classroom volume meter earlier. This script checks a sensor constantly, and changes between behaviors based on the last seen value.

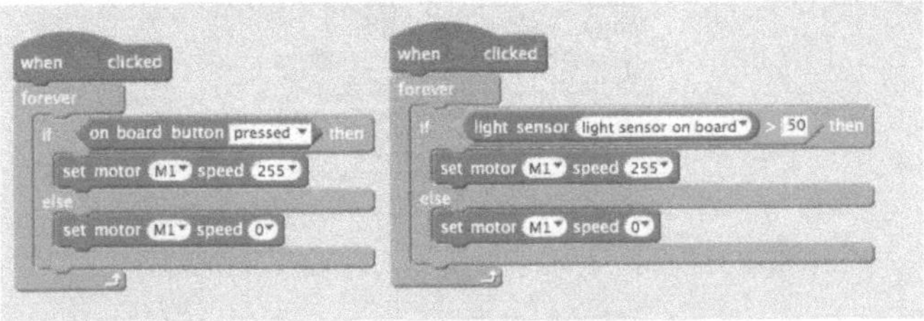

While a digital sensor only swaps between two values, analog sensors generate ranges of values and behaviors. When you're looking for a program to do "this, or that, or that other thing," it's time to expand the state check script. This variation of the state monitor uses nested If/Else blocks.

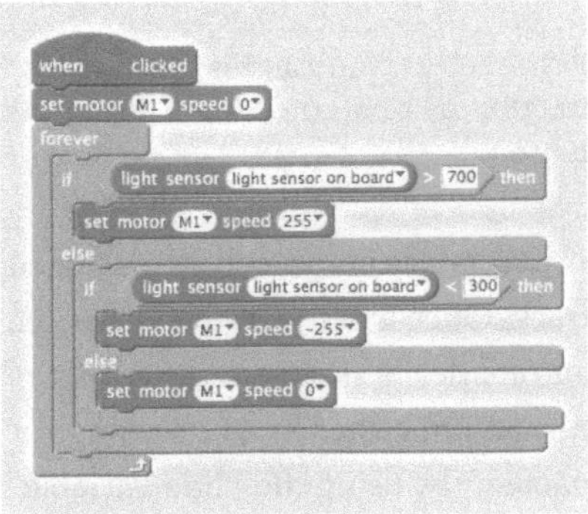

It's important to note that this script reads the sensor twice. Since these readings happen in quick succession, it's reasonable to assume that the light levels haven't changed drastically, but they can generate different values. Furthermore, when mBlock programs are running tethered, each sensor reading requires two-way communication between the computer and the mCore board. This communication should take less than 100 milliseconds, but it might take longer, and that delay will only grow as the program becomes more complex.

To avoid these problems, we need some way to store a sensor reading and check it several times. In other words, we need a variable.

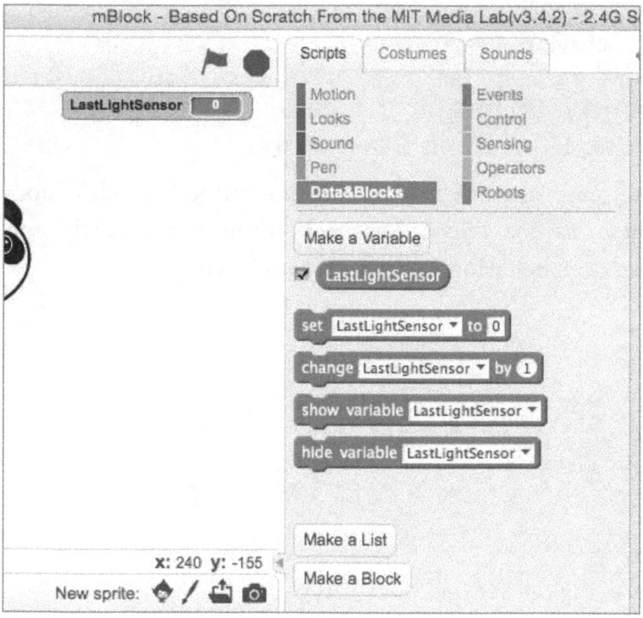

FIGURE 2-21: You can find the blocks for variables and lists in the Data&Blocks palette. Select either the Make a Variable or Make a List button.

Variables in mBlock are designed to be approachable and easy to track. Once a new variable is created and named (name it well!), it's automatically shown in the corner of the screen. This display can be turned off, either by unchecking the small box next to the variable name or by using the Hide Variable block.

There are only a few variable blocks . The block Set *VariableName* does just that—overwriting any current data and leaving the new value. The block Change *VariableName* increments, or decrements with a negative number, the current value. Most importantly for our current recipe, the oval reporter for *VariableName* can be used all over mBlock, in any round input spot.

Here's that three-state check from Figure 2-21 rewritten to use a variable to store the Light Sensor value.

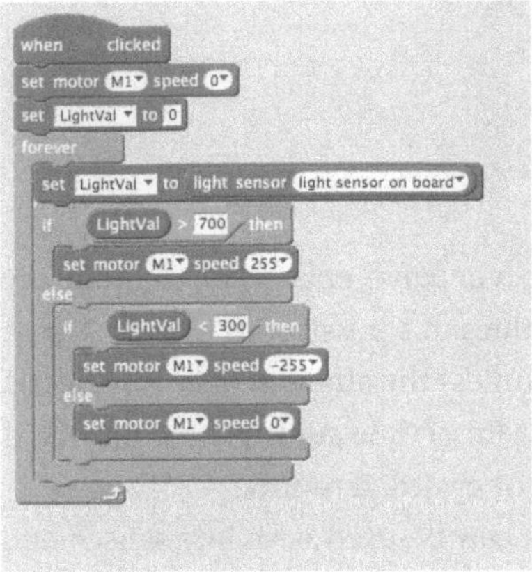

Now the Light Sensor is read once, at the top of the loop, and the value is stored in LightVal. All the checks are made against this stored data, rather than reaching out to the sensor itself. The loop is now protected from sudden changes in the sensor data, and the communication time between mBlock and the mCore board is minimized.

This loop checks the sensor reading against two threshold values, giving three possible outcomes: low, high, and between. In this example, the motor runs forward at full speed if the value in LightVal is above 700, runs backward full speed if it's below 300, and turns off for any values between 700 and 300.

Using the mBlocks `<AND>` operator, a single `<IF>` statement can check for values between two thresholds. With this technique, we can slice a sensor reading into a large number of discrete segments. Since the `<IF>` statements are constructed so that only one can be true in a given moment, these are called *switch cases*.

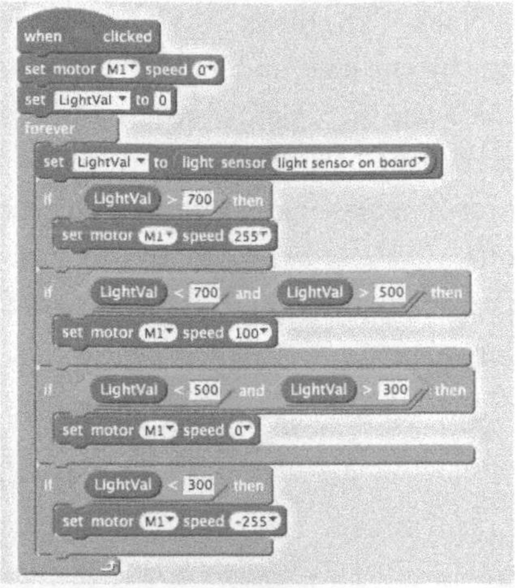

The performance of switch cases depends entirely on the accuracy of the sensor and the threshold values. It's possible to write a 12-part switch case for a light sensor, but unless the ambient light is perfectly consistent, you will have to spend a lot of time adjusting the threshold values to account for cloudy days or crowded rooms.

Remember that these recipes can be used with any sensor and any output behavior. If the goal is to turn on a specific number of lights or perform other discrete actions, then switch cases like these are a dependable tool.

"When the Value on the Sensor Grows, Do More Stuff," Also Known as *Proportional Control*

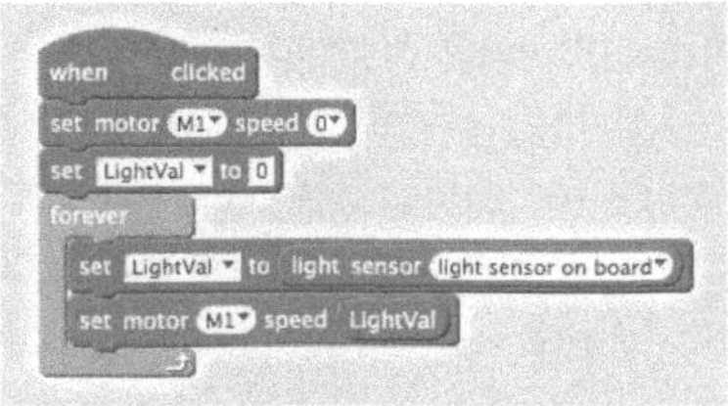

In the following code, LightVal is tied to the numeric value of the light sensor and is used to directly control the speed of the motor. This sounds great at first—when the light is dim the motor speed will be low, and when the light is bright the motor speed will be high, right? The reality will be a bit underwhelming. When there's a gap between concept and execution, it helps to look for the assumptions in the program. By using the value in LightVal as the motor speed, our program assumes that the sensor generates values exactly within the motor's input range.

The Me Light Sensor reports values in a range from approximately 0 to 1000, where comfortable indoor lighting ranges between 400 and 600. The M1 motor block can spin in either direction, with -255 being full reverse and 255 full speed ahead. Additionally, speeds too close to zero don't generate enough force to turn the gears in the yellow mBlock motors. This mismatch in sensor output and motor input values explains the dull behavior in the loop above. Directly plugging the light sensor values into the motor block will turn it full speed forward in most lit rooms. Worse, since the light level is never negative, the motor will never spin in reverse.

With a particular sensor, a given output, and a little time, it's easy enough to throw together some arithmetic to squash the sensor values into the input's ideal range. If the light values in our room are between 400 and 600, but we want motor speeds between -255 and 255, we could use subtraction to shift the range.

The act of translating a value from one range to another is called *mapping*, and we can turn this mapping into a custom block.

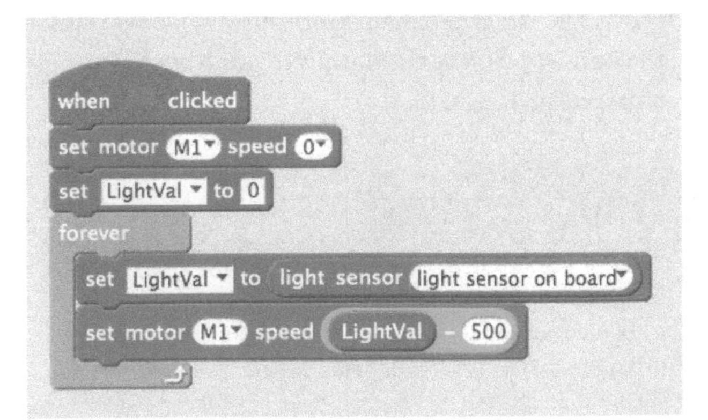

FIGURE 2-22: If ambient light readings range between 300 and 700, subtracting 500 from each reading will generate motor outputs in the -200 to 200 range, causing erratic back-and-forth motions.

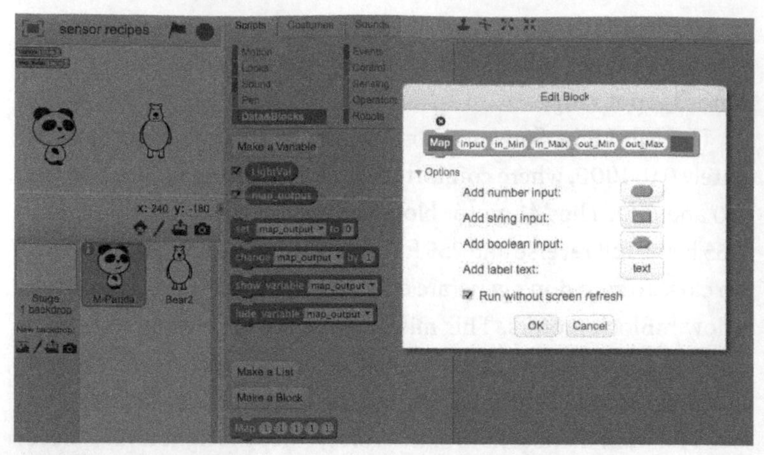

FIGURE 2-23: Custom blocks in mBlock are tied to particular sprites. Here, the panda knows how to use Map, but the polar bear doesn't.

Custom blocks are powerful tools that can do wonders for the readability of mBlock programs. As a general rule, if a particular sequence of blocks shows up more than twice in a program, they should probably collected as a custom block.

It's important to note that these custom blocks are not distinct functions or programs. They share namespace with, and can perform the same actions as, the larger program. Since custom blocks cannot be reporters, the best way to save the output of our Map function is to create a variable called `map_output`.

```
define Map input in_Min in_Max out_Min out_Max
set map_output to input - in_Min * out_Max - out_Min / in_Max - in_Min + out_Min
```

FIGURE 2-24: This custom block re-creates Arduino's `map()` function. For more information on that function, see *https://www.arduino.cc/en/Reference/Map*.

To use this version of Map, we need to know the range of the input and the range of the output. Once that's settled, we can drop the sensor block into the first bubble.

```
when clicked
set motor M1 speed 0
set LightVal to 0
forever
  Map light sensor light sensor on board 0 1000 -255 255
  set LightVal to map_output
  set motor M1 speed LightVal
```

FIGURE 2-25: Place the value you want to shift in the first Map slot, followed by the range of that input, then the desired output range.

FIGURE 2-26: The Me Sound Sensor reports analog values and must use the mCore's analog inputs, either port 3 or 4.

When we connect a new sensor, our first test program is always a Say block in a Forever loop. This tests the hardware and software connections all the way from the sensor to mBlock's Stage, and shows real-world sensor values. With this tiny script running, test out the Me Sound Sensor. Scream for a bit. Type on the keyboard, and bang on the table. Have everyone in the room hold their breath for 10 seconds. While the technical sensor values might range from 0 to 1024, it's far more valuable to see what values *your* quiet and *your* loud generate in your particular room. Only the actual data from your environment will create useful threshold values for your program. The numbers used in the following traffic light examples are tied to the classroom and kids who designed these particular traffic lights.

Watching the volume data flicker in a Say block's word balloon is a good reminder that sound levels in a room change rapidly. Volume measures a constant sequence of momentary noises: a dropped book, squeaking chair, or collective breath. As someone prone to making loud sneezes, I'd like this traffic light to ignore some momentary volume spikes and respond instead to steady increases over time. To do this, we'll open the door to the world of sampling.

The block in Figure 2-27 shows a very basic way to sample a sensor and report back the mean value instead of a single reading. Here, this means creating one more variable, named RecentSounds in the example, and using it to store 10 sensor readings. Note that RecentSounds doesn't keep 10 distinct readings; it just adds all of the values together. Using lists instead of variables, mBlock can store persistent collections

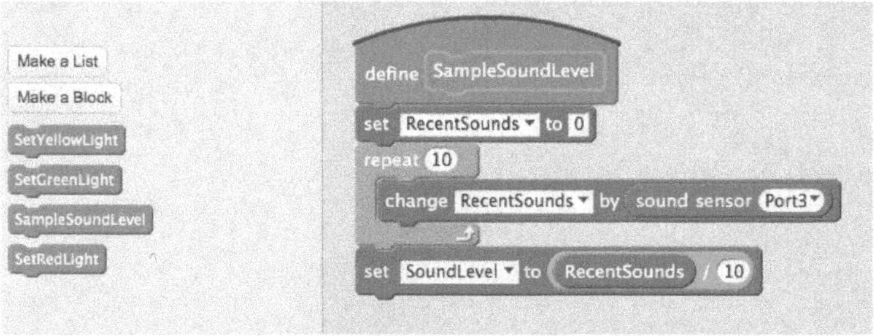

FIGURE 2-27: Using custom blocks allows us to visually hide the complexity of this sampling process outside the main body of our traffic light program.

of incoming data, which we explore in Chapter 4. After the readings are complete, the average is stored in the familiar SoundLevel block. There's no Forever loop in this block because it's designed to be used as a single command in a larger program.

So far, we've focused on monitoring the sound levels in the room using the microphone sensor. To turn that passive sensor into a traffic light, we'll need to dive into programmable RGB LEDs. Traffic lights in the real world don't normally use color-changing lights—the top is always red, and the bottom is always green. While the mBot can power a bunch of fixed-color LEDs, the LED accessories they sell are all *addressable*, meaning that every light in the strip can have a unique color. These RGB lights are technically WS2812, similar to Adafruit's original NeoPixels. Makeblock sells several programmable RGB LEDs in several different forms, but they all work the same way in mBlock and the Makeblock app.

Connect each light board to a different port on the mBlock. Since the Me Sound Sensor is currently using port 3, this means connecting a light to ports 1, 2, and 4.

Using a separate block for each light source (either the Me LED board or the longer Me LED strips), specify the intensity for the red, green, and blue channels with a value between 0 and 255.

set led Port4 all red 255 green 0 blue 0

There are plenty of fancy tricks available for working with these lights in mBlock, but for the traffic light all we need is to have one light be the appropriate color and have the other two off.

set led Port1 all red 0 green 0 blue 0
set led Port2 all red 0 green 0 blue 0
set led Port4 all red 255 green 0 blue 0

This program is human-readable up to a point. When the mBlock Scripts panel is filled with these nearly identical blocks, it's really easy to lose track of whether port 2 is supposed to be the green light or the yellow light. To make the script easier to parse, we can create a custom block—this time to simply isolate the LED blocks into meaningful, named groups.

This revised program now combines the screen-based prototype with real-world lights, and now students have a great opportunity to cross-check behaviors between the two.

UPLOAD TO ARDUINO

All of the projects so far haven't actually changed the bits written into the memory of the mCore. Using the remote control to move the mBot around doesn't change the software. The Makeblock app and the programming we've done so far in mBlock constantly send commands to the mBot, but never rewrite the program stored in memory on the board.

Now, we'll move from tethered to independent operation of the mBot. Using a wired USB connection, we can upload a program directly to the mCore that will stay loaded through resets and power

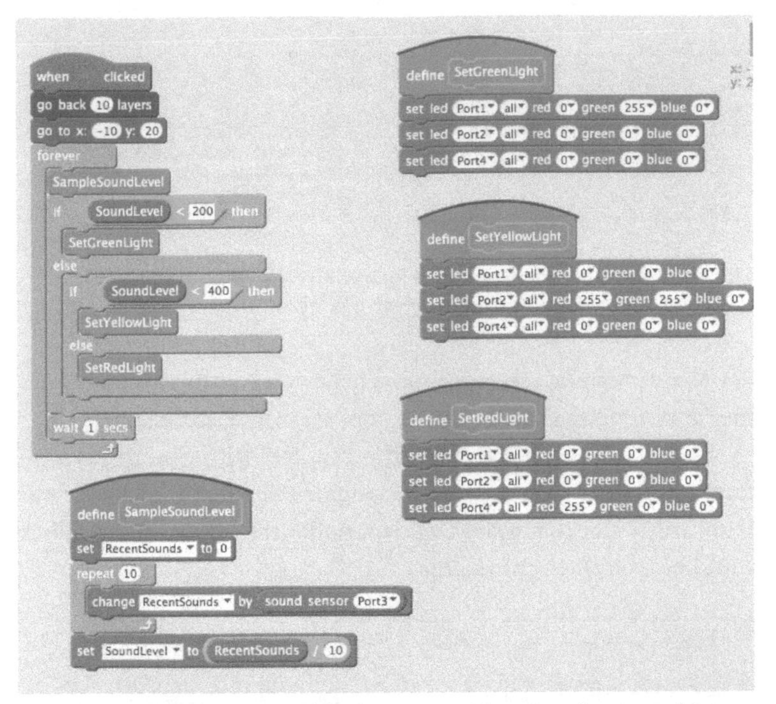

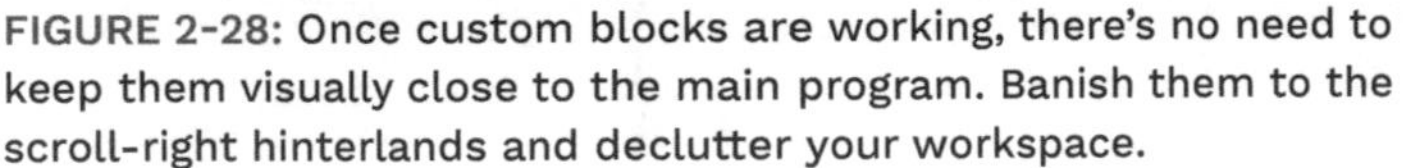
FIGURE 2-28: Once custom blocks are working, there's no need to keep them visually close to the main program. Banish them to the scroll-right hinterlands and declutter your workspace.

cycles. Uploading a program is the only way to create a robot that that operates without a computer on hand.

At this point, anyone familiar with "normal" Arduino or microcontrollers is sighing with exasperation, "Finally!" Uploading code to the board to blink an LED is how 99 percent of Arduino tutorials start. Tethered programs have provided a ton of features that disappear when uploading code to the mCore. Untethered, there's no interaction between the board and the computer, hence no way to use many of the mBlock features derived from Scratch. Every block you use will be translated into written Arduino code. When you compile and upload an mBlock program, it needs to have a different "hat" than a tethered program. (Scratch convention names the curvy, top-of-the-script block a *hat*.)

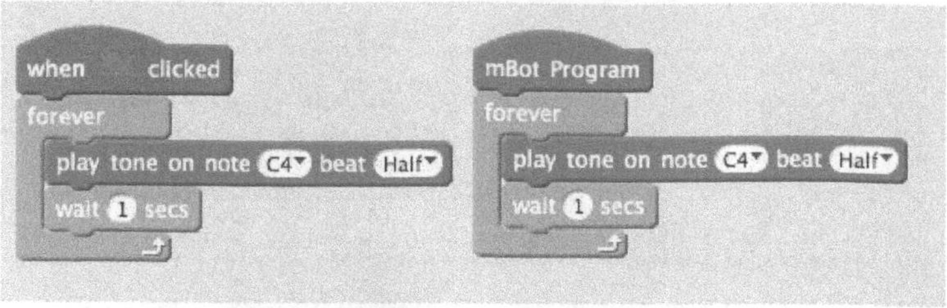

FIGURE 2-29: Only the GreenFlag hat will work when the mBot is tethered, and the mBot Program hat has no effect.

These programs create the same behavior, but the script that uses the mBot Program hat requires compilation and upload.

In general, only blocks from the Data&Blocks, Control, Operators, and Robots palettes work in compiled and uploaded programs. If any other blocks appear in a script under the mBlock hat, mBlock will show you an error message.

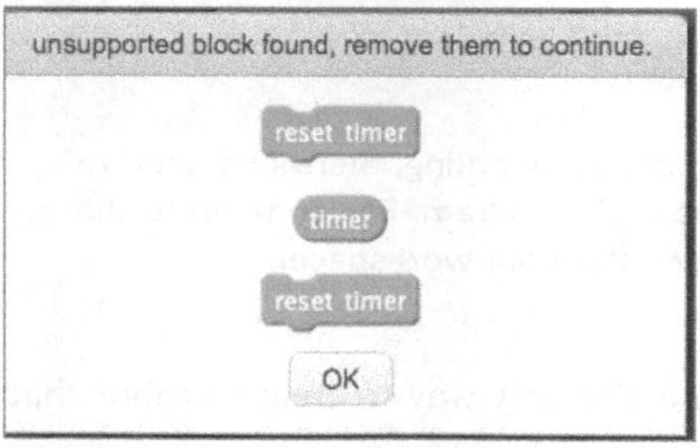

FIGURE 2-30: These light blue blocks come from mBlock's Sensing palette and control timing functions inside mBlock. They can't be used in an uploaded program.

INDEPENDENT TRAFFIC LIGHT CLASSROOM VOLUME METER

If the volume meter is ever going to be useful in a classroom setting, it needs to work like an appliance—flip the power switch and the lights start right up. Fortunately, mBlock makes it simple to change an interactive program that requires a computer to one that's uploaded to the board and works independently.

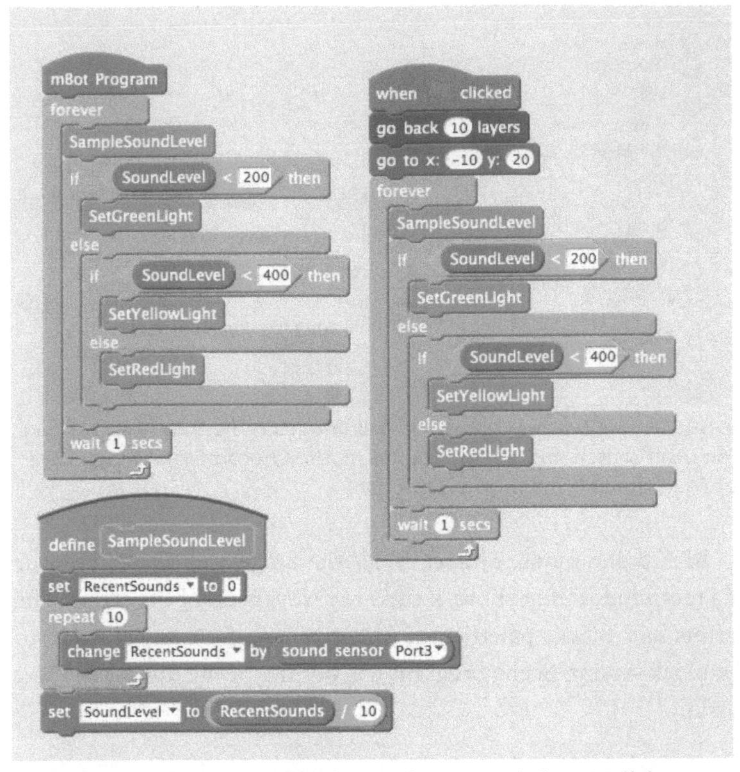

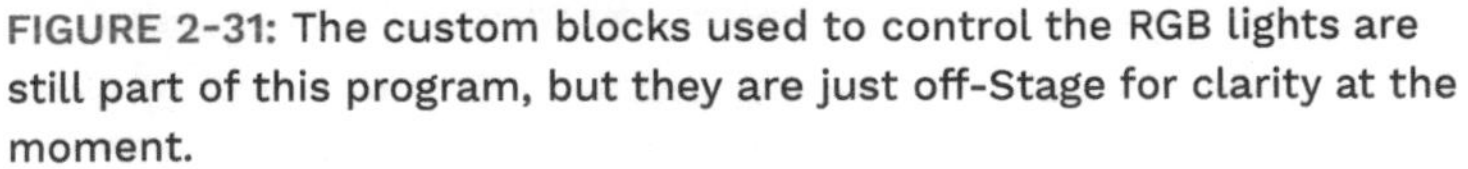

FIGURE 2-31: The custom blocks used to control the RGB lights are still part of this program, but they are just off-Stage for clarity at the moment.

Nothing from the earlier version of the program has been removed. We've added an extra script under the mBot Program hat and removed the two blocks that referred to the sprite's position on the screen. When we compile and upload, only the blocks under the mBot Program hat and any custom blocks used in that script will translate into Arduino code. This means that the tethered version can coexist with the compiled version in one mBlock file. If the green flag is clicked while an mBot is connected, the tethered program will run. If you want to have the program run without a computer, select Arduino mode by right-clicking the mBot Program hat. You can also go to the Edit menu and select it there.

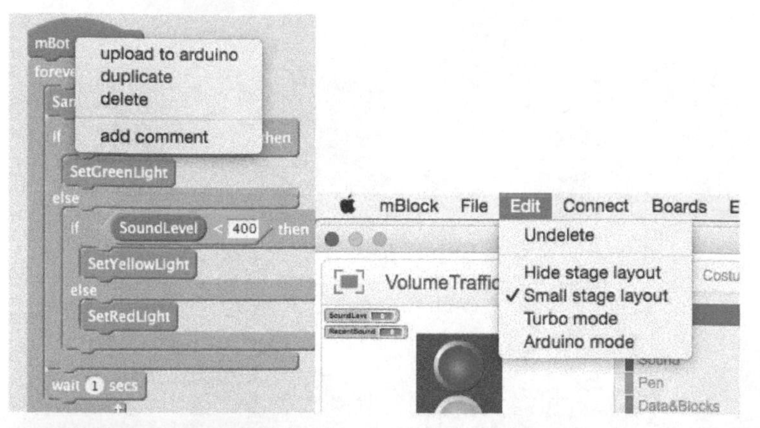

FIGURE 2-32: Choosing Upload to Arduino from the mBot hat context menu will switch mBlock to Arduino mode. Choose Small Stage Layout from the Edit menu to switch back.

In Arduino mode, mBlock hides the Stage and sprites in favor of a text window that shows a current text version of the script. The Scripts and Blocks palettes are still visible, and changes made to the block version of the program will automatically update the text version.

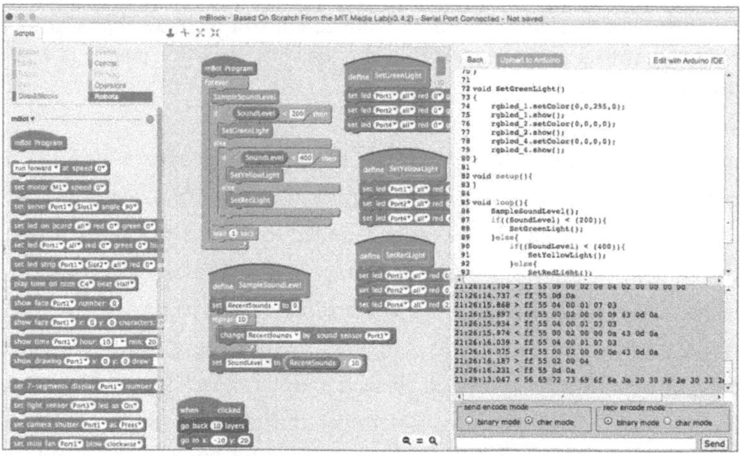

FIGURE 2-33: On the lower-right corner of the screen when you're in Arduino mode, you can see output from the compiler, and data that would be sent to the serial monitor in the Arduino serial monitor when using traditional Arduino tools.

Careful reading of the text version can reveal a lot about how mBlock translated between blocks and Arduino code. If you look closely at the text version on the right of the following image, you'll see that the SetGreenLight custom block has arrived as the Arduino function `void SetGreenLight()`.

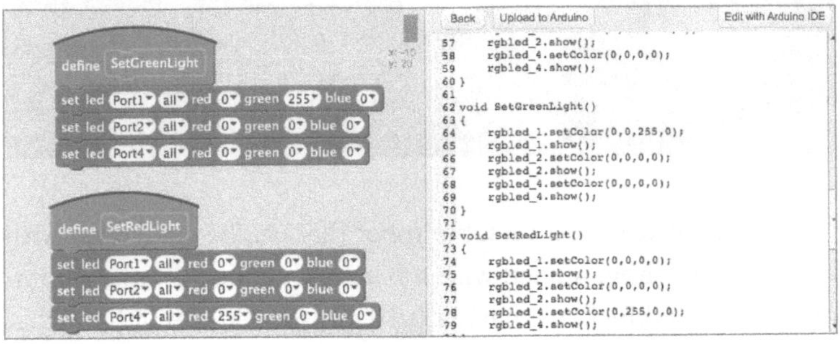
define SetGreenLight
set led Port1 all red 0 green 255 blue 0
set led Port2 all red 0 green 0 blue 0
set led Port4 all red 0 green 0 blue 0
define SetRedLight
set led Port1 all red 0 green 0 blue 0
set led Port2 all red 0 green 0 blue 0
set led Port4 all red 255 green 0 blue 0
Back
Upload to Arduino
Edit with Arduino IDE
57 rgbled_2.show();
58 rgbled_4.setColor(0,0,0,0);
59 rgbled_4.show();
60 }
61
62 void SetGreenLight()
63 {
64 rgbled_1.setColor(0,0,255,0);
65 rgbled_1.show();
66 rgbled_2.setColor(0,0,0,0);
67 rgbled_2.show();
68 rgbled_4.setColor(0,0,0,0);
69 rgbled_4.show();
70 }
71
72 void SetRedLight()
73 {
74 rgbled_1.setColor(0,0,0,0);
75 rgbled_1.show();
76 rgbled_2.setColor(0,0,0,0);
77 rgbled_2.show();
78 rgbled_4.setColor(0,255,0,0);
79 rgbled_4.show();

Selecting Upload to Arduino will launch the compiler. The compiler will translate the human-readable Arduino program into a hex file, and then will upload that hex to the mCore. Error messages that appear in the lower-right window during compilation are often a weird combination of compiler errors and serial communication codes. Troubleshooting that universe of errors is well beyond the scope of this book. In practice, most errors that students encounter at this stage can be traced back to mBots with disconnected USB cables. If this is the first time you're using a particular computer to upload a program, make sure the Arduino drivers are installed. That process is covered earlier in this chapter in the overview of wireless, Bluetooth, and USB connections.

Once the upload is complete, the traffic light volume meter program is now written to the mCore's stable memory. Turn off the board, unplug the USB cable, and build a better traffic light.

Finally! Instead of a versatile robot that can be controlled with an IR remote, programmed with a Bluetooth tablet, or issued commands from mBlock, we have a battery-powered traffic light that responds to noise. Moving from the flexible tool that could become anything to a narrow, single-purpose *thing* is a huge step for young designers. But it shouldn't be the last step in the design process.

Half-built cardboard prototypes like this often represent the endpoint of student projects. From the perspective of a student following a strict feature checklist, this traffic light volume meter is clearly "done." Making any changes will involve undoing something that already works, an idea that is anathema to goal-focused learners. Without stepping on the learners' celebration, we insist on reflection and peer review at these seemingly terminal prototype stages. Young Makers need to develop an iterative mindset and an eye for improvements, even when it means "redoing" work. One of the best ways to force this reflection is to put the prototype in use, and have testers deliver honest feedback to the designer.

Simple critiques can prompt significant changes in the design. If a user wants a way to adjust the volume levels on his or her own, how many systems does that effect? First, it means adding some extra form of input to the project, probably some buttons or a potentiometer. As a result, that means using fewer ports for the lights. Is it better to use the LED strip, or position the mCore to use the onboard

LEDs? Which design would be more stable and allow teachers to place the light vertically or horizontally?

The mBot's expandability, combined with mBlock's beginner-friendly programming syntax, makes it easy to start creating interactive projects like the traffic light. But the challenges of design come from refining those initial prototypes into something that meets the demands of real-world users and environments. Use the power and convenience of these tools as a shortcut to those hard/fun problems.

REINSTALL THE DEFAULT PROGRAM

As hard as it might be to believe, the novelty of this stand-alone Ssssh-meter will wear off. When that happens, you'll need to replace the traffic light volume meter program with a program that can communicate with mBlock.

Use a USB cable to connect the mCore to the computer and connect to mBlock via the serial port. Then select between the two confusingly named mCore options: Update Firmware and Reset Default Program.

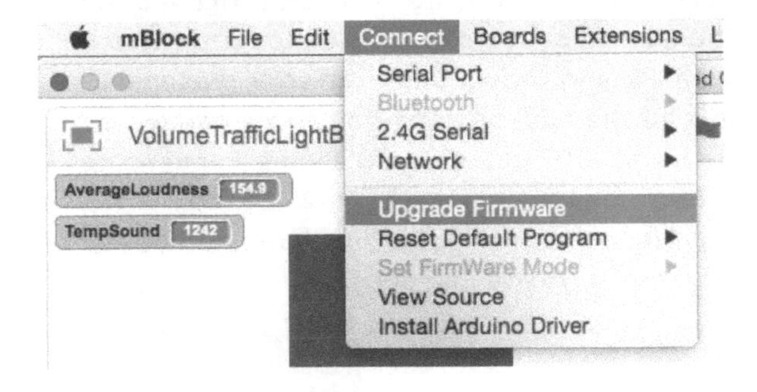

There is a detailed comparison between these two versions in Chapter 1, "From Kit to Classroom," but the takeaway is that Update Firmware is the better choice unless you're planning on using the IR remote. Make your choice, and then wait for the three-tone chime (Default Program) or chirp (Update Firmware) when the upload is complete.

WHERE WE'RE HEADING FROM HERE

With this wide array of tools, it's clear that "How do I control my mBot?" is the wrong question. Given any task, there's probably a way to accomplish it using the Makeblock app or mBlock, or by using the Arduino environment. For an open platform like the mBot, you can choose the most focused tool, or the most flexible tool, or just use one with which you're already comfortable. The remaining project chapters will each use one specific software tool, mainly for clarity in the instructions. We'll call out any unique features of a particular programming environment when we use them. Other than those exceptions, you should be able to build all the animatronics and data loggers from the following chapters using the Makeblock app, mBlock, or the Arduino IDE.

Sensors and Example Code

Onboard sensors are the sensors built right into the mCore, the brains of the mBot. Two sensors, the ultrasonic sensor and line-follower, are included with the basic mBot kit. The add-on sensors are available for purchase in bundled packs and individually for very reasonable prices. Nearly all the add-on sensors can be connected to the mCore using RJ25 (phone jack) cables. For sensors that are not made by Makeblock, the RJ25 adapter listed here is the perfect solution. The RJ25 adapter allows you to connect your own servos and sensors.

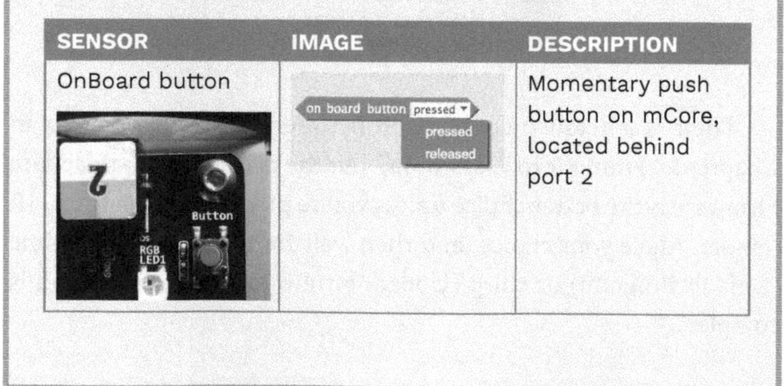

SENSOR	IMAGE	DESCRIPTION
OnBoard button		Momentary push button on mCore, located behind port 2

SENSOR	IMAGE	DESCRIPTION
OnBoard LED x 2	set led on board all red 0 green 0 blue 0	Two programmable RGB LEDs, located between port 2 and port 3
OnBoard Light sensor	light sensor light sensor on board	Wide-angle analog light sensor, mounted directly between the onboard RGB lights
OnBoard buzzer	play tone on note C4 beat Half	A standard Piezo buzzer; the Play Tone block allows notes from C2 (==65 Hz) to D8 (4700 Hz), in half to double duration

SENSOR	IMAGE	DESCRIPTION
OnBoard IR sensor	mBot's message received ir remote A pressed	IR receiver and transmitter are mounted next to each other, between the speaker and the button
Ultrasonic sensor (Included with mBot kit)	ultrasonic sensor Port3 distance	Measures distance from 3 cm to 400 cm and can be used for obstacle avoidance and measuring distance
Line Follower sensor (Included with mBot kit)	line follower Port2 line follower Port2 leftSide is black	Two LEDs and light sensors mounted to a single board. This sensor is calibrated for the height of the mBot frame. Be sure to test when using in other situations.
Add-on Me LED 4x	set led Port1 all red 0 green 0 blue 0	Four RGB LEDs that can be adjusted for color and brightness

SENSOR	IMAGE	DESCRIPTION
Add-on 7-segment display	set 7-segments display Port1 number 100	Can be used to display data such as speed, time, temperature, distance, or a score
Add-on Sound sensor	sound sensor Port3	Electret microphone. Detects loudness of sound at close range.
Add-on Potentiometer	potentiometer Port3	Can be used to adjust speed and brightness of objects
Add-on PIR Motion sensor	pir motion sensor Port2	Detects motion of humans or animals in a 6 meter range
Add-on Joystick	joystick Port3 X-Axis X-Axis Y-Axis	Used to control the direction of physical objects or video games
Add-on Light sensor	set light sensor Port3 led as On light sensor light sensor on board light sensor on board Port3 Port4	Detects the intensity of ambient light

SENSOR	IMAGE	DESCRIPTION
Add-on LED matrix		8×16 aligned LEDs to display numbers and letters
Add-on RJ25 Adapter	n/a	Converts standard RJ25 to six pins to use generic servos and sensors
Add-on LED Strips		WS2812 programmable LED. Uses an RJ25 adapter to connect to mCore.
Add-on Temperature Sensor		Measure inside or outside range between –55°C and 125°C. Sensor is waterproof and uses RJ25 adapter to connect to mCore.

3

Animatronics

Every kid wants to build a robot. No matter what materials are at hand, from cardboard to empty soda bottles to brooms, if a kid starts to build, there's a decent chance that the shape that emerges will be named *robot*. With that type of enthusiasm and access to real, powerful components, the perfect robot should emerge spontaneously, right?

Using context to create focus is a key to any successful work with young or inexperienced roboticists. Left to describe their dream robot, most kids will describe some fantastical blend of Baymax, Optimus Prime, and Gundam Wing. Vision that expansive can inhibit, rather than inspire, when it hits the hard reality of servo motors.

This group of projects focuses kids' attention on a "simple" branch of robots that move and respond to the environment for the benefit and enjoyment of an audience. Kids can think of these as interactive tops, preprogrammed puppets, or scaled-down versions of audio-animatronics developed by Walt Disney Imagineering. While working on the Mission to Mars ride at Disneyland during college, Rick got firsthand experience with Disney's audio-animatronics brilliance. Both vintage and newer Disneyland rides include this trademarked Disney technology.

First, we'll build some puppets that make random movements to introduce several different operations, and then move on to more advanced creations that actually respond to user input. Each section will explain the specific hardware needed for movements and sensing. Having a handful of custom-made RJ25 cables using the instructions in Chapter 1, "Kit to Classroom," will be very handy for these projects. The short 6″ cables that come with the mBots will seriously limit your creativity. With cables 1′–3′ long, you can really accomplish almost anything you dream up. For all the projects in this chapter, the box-creature bodies are just a starting point and will surely turn into whimsical creatures as kids' imaginations go wild.

Materials

TOOLS

- Hobby knife
- Masking tape
- Cutting mat
- Sharpie
- Pencil
- Hot glue gun and glue sticks
- Scissors
- Needle-nose pliers
- Ruler

CRAFT SUPPLIES

- Boxes of various sizes
- Rubber bands, small and large
- Feathers
- Pipe cleaners
- Beads
- Bling
- Paint
- Colorful foam sheets
- Craft sticks, jumbo and regular
- Paper or plastic cups
- Googly eyes
- Colorful construction paper
- Large paper clips

ELECTRONICS

- mCore (preferably with case)
- Sensors and motors (see list that follows)
- RJ25 connection cables (You'll want to use the instructions from Chapter 2 to make your own, since you'll want longer ones than are in the kit.)
- RJ25 adapter (for using generic servos)

FOR ADDING SENSING AND MOVEMENT

Sensors to trigger (input)

- Ultrasonic
- Distance
- Motion
- Light
- Line-following
- Sound
- Touch

Components to react (output)

- Servos and linkage arms (aka, servo horns)
- LEDs
- Motors

PUPPET MOVEMENT WITHOUT SENSORS

For the first few projects in this chapter, we'll build some creations that light up, rotate, and spin, but don't react to user input. Later on, we'll build some things that actually respond to user input using specific sensors.

Project: Random Light-up Eyes Using RGB LED Sensor

In this first project we'll create a basic cardboard box head with cut-out eyes and an RGB LED inside.

1. Select a box that is approximately 5″ × 5″ × 5″—I used an empty tissue box. Open the box so you can get inside.

2. Cut some eye holes out with a hobby knife and then add

some tissue paper on the inside to cover the holes and diffuse the light.

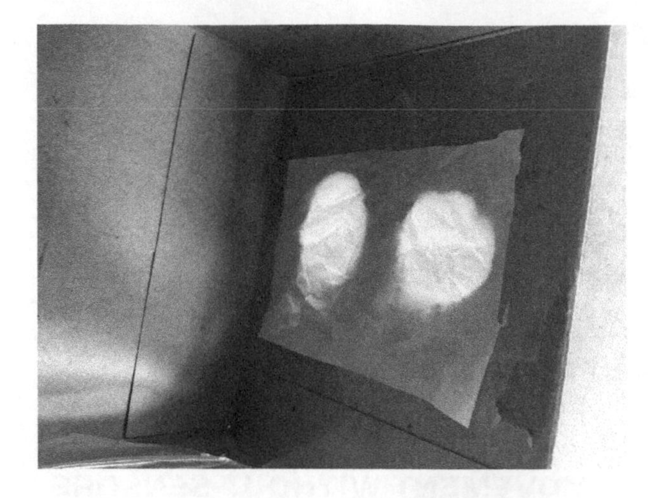

3. Put masking tape on the bottom part of the LED sensor. (Whenever you're going to use hot glue on a sensor, add tape first to prevent damage to the electronic parts.)

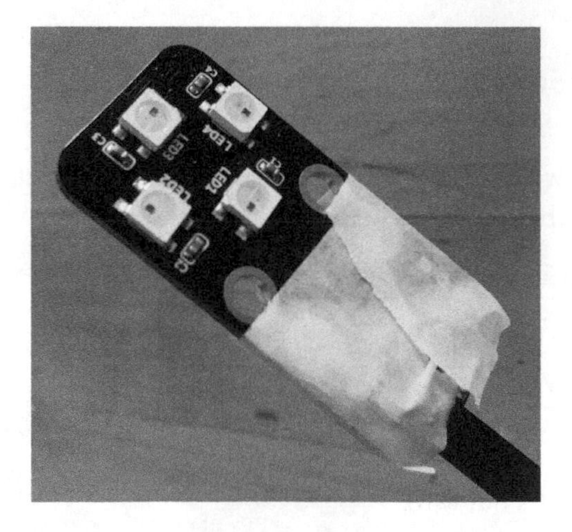

4. Add hot glue to the tape on the LED sensor and stick it inside the box.

5. Thread the RJ25 cable out the bottom of the box and attach the cable to port 1 on the mCore.

6. Connect the mCore to your computer and open mBlock. Write and run the following program:

```
when space key pressed
forever
    set led Port1 all red 0 green 0 blue 255
    wait pick random 1 to 10 secs
    set led Port1 all red 0 green 0 blue 0
    wait 2 secs
```

This will create randomly flashing blue LED eyes that run forever after the space bar is pressed. This is just a starting point, so now it's time to get creative by modifying this code to customize the colors and blinking patterns.

Project: Head Turning Randomly Using 9g Servo and RJ25 Adapter

If you are using a lightweight "head" like the tissue box for this project, 9g servos will work, with some modifications. If you are moving a heavier object, you might need a bigger servo, like a Hitec HS-311, which has a higher torque. For $3–$5, you can also purchase micro servos with metal gears that are less likely to be stripped by too much weight or force.

Mount and Wire the Servo

1. First, cut a mount for the servo. If you have access to a laser cutter, download the template file from *www.airrocketworks.com/instructions/make-mBots*, and use it to cut a mount from acrylic. If you don't have access to a laser cutter, you can also use the full-scale PDF, which can be downloaded from the same location, to hand-cut a servo mount out of a material of your choice. Slip the servo through the mount and attach it using hot glue.

2. Next, cut a hole about ¾″ × 1¼″ in the center of the box to fit the servo and push the servo up through the hole in the box.

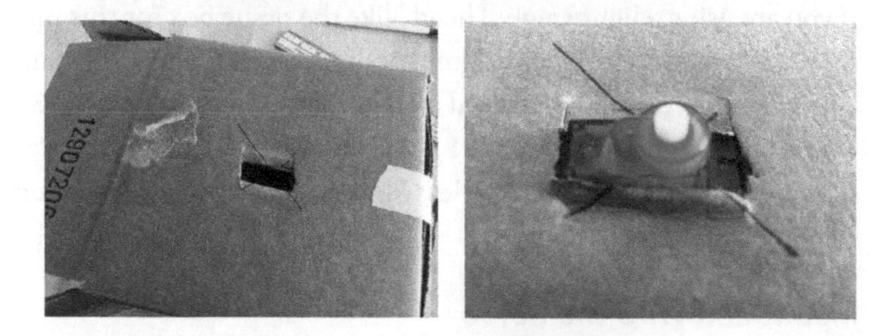

3. Once you're sure the servo fits, put tape over the servo mount and glue it to the inside of the top of the box.

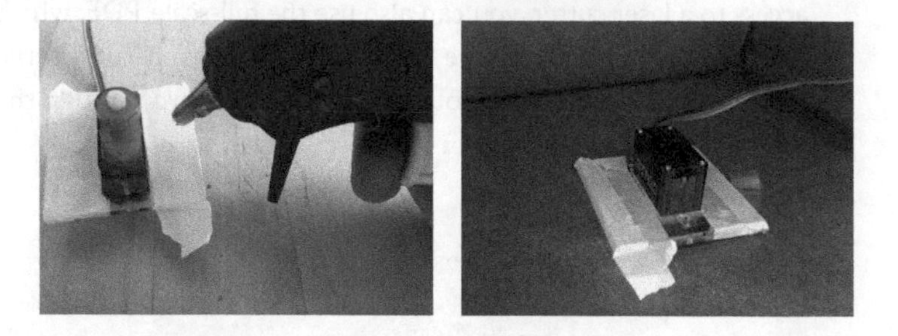

4. Feed the servo wires out the back of the box and tape the box shut. Then, connect the servo to the Makeblock RJ25 adapter. The RJ25 adapter allows you to connect two servos to one port on the mCore. For this project, let's attach the servo to slot 1 using the following guidelines:

   » Orange or yellow: S1 (signal)

   » Red: VCC (power)

   » Brown or black: GND (ground)

BUILDING THE SERVO ARM

1. Mark the center of a large craft stick and drill a ¼″ hole.

2. On a piece of cardboard, trace whatever you are using for your creature's head, and then cut out the shape. I'm using the same box I used for the LED eyes project because I'm going to use this as my creature's head.

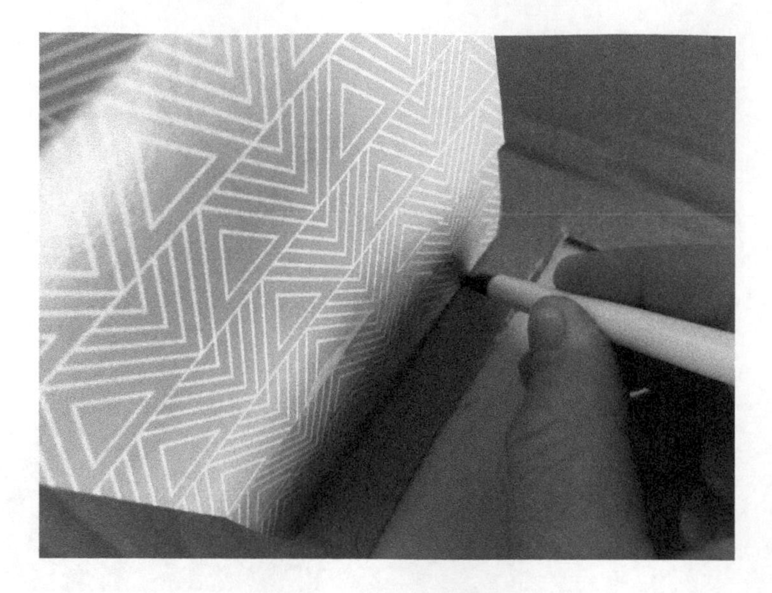

3. Hot-glue the craft stick to the piece of cardboard, then drill through the hole in the stick again and on through the cardboard. Depending on the size of the box used for the head, you may need to trim the craft stick.

4. Next, take the largest arm that came with your servo and hot-glue it onto the stick, with the side that attaches to the servo facing up. You need this to stick really well, so use plenty of glue, but not so much that it goes inside the hole.

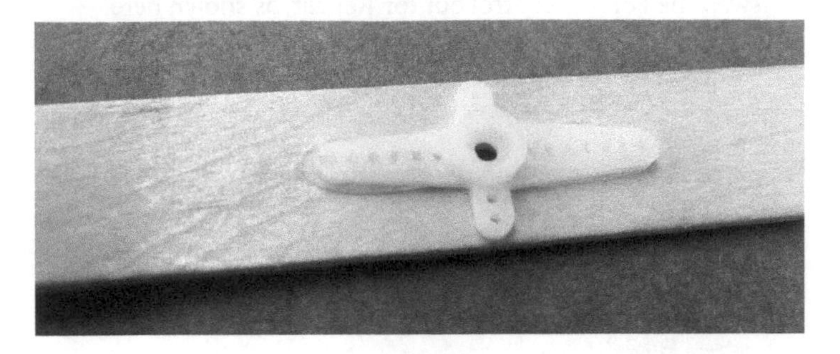

We'll set this aside for now until we have the servo connected to the mCore and calibrated.

WIRING TO THE MCORE

1. The servo should already be connected to the RJ25 adapter. Now connect the RJ25 adapter to port 2 on the mCore using an RJ25 cable. (It's connected and programmed for port 2 because you may want to use port 1 for the LED eyes.)

2. Connect your mCore to your computer and write the code shown in the following image using mBlock. This will make the servo turn from 0 to 180 degrees randomly when the space key is pressed. If you want the servo to move for only a set period of time, you can swap the Forever control out for Repeat, as shown here.

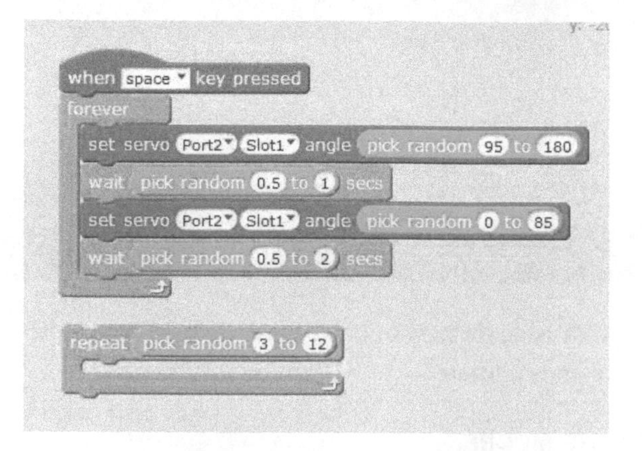

ATTACH THE HEAD TO THE BODY

1. Once the mCore is connected and programmed, attach the stick and cardboard control arm to the top of the servo and carefully screw it in place using the screw supplied with the servo.

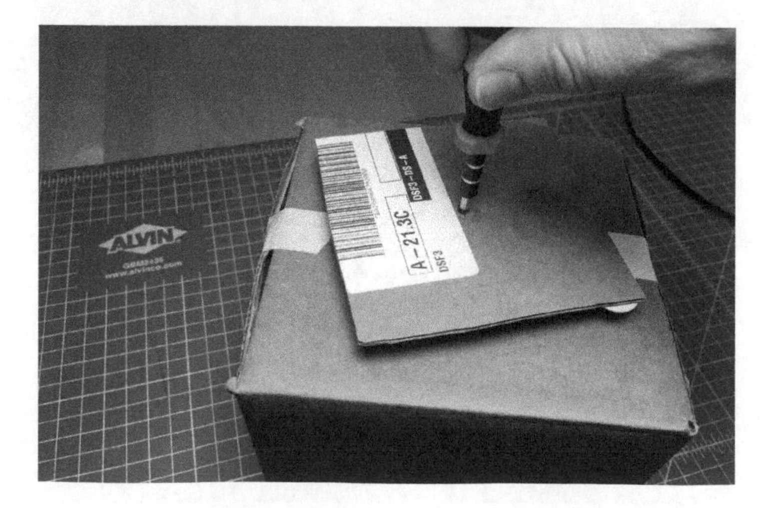

2. Then, attach the head to the servo platform using tape.

COMBINING THE LED EYES WITH THE MOVING HEAD

1. If your LED eyes are still inside the head, you'll need to route the cable out and to the mCore in a way that the cable does not interfere with the operation of the moving head.
2. Now, using one of your custom RJ25 cables that's at least 2′ long plug your LED eyes into port 1 of the mCore and combine the two programs. Now you've got a randomly moving head and blinking LED eyes!

Going to the Zoo

The Robot Petting Zoo was born out of the TechHive Studio at the Lawrence Hall of Science at the University of California, Berkeley. The first Robot Petting Zoo was a brilliant 12-hour Makeathon for high school students. During the first 10 hours, students learned about programming, electronics, prototyping, and design while they built a robot pet, and then spent the last two hours presenting their creations to the public. The Robot Petting Zoo was the inspiration for some of the following projects. Matt Chilbert and Andrew Milne were instrumental in the event at TechHive, and Tom Lauwers is the president and chief roboticist at BirdBrain Technologies. Thanks to these gentlemen for their inspiration!

Project: Opening Mouth Using a 9g Servo and an RJ25 Adapter

This puppet in this project is made using a box, one servo, a single-sided servo arm, a small craft stick, and a large paper clip to operate the mouth flap.

We're going to start with a cardboard box and use one of the flaps as the mouth. For this example, we're using a 6″ × 6″ × 6″ box from Uline.

1. Lay the box flat and, on one end, cut off the two flaps opposite of each other.
2. On the side you didn't cut, tape the four flaps closed with a strip of masking tape.

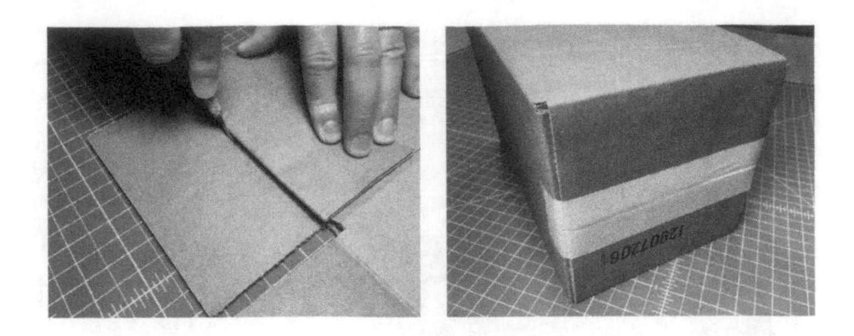

3. On the other side, tape just one flap down and draw eyes. The bottom flap will be our mouth.

4. Grab a 9g servo and select the longest single-sided servo arm. We'll be extending the arm by hot-gluing a craft stick to it. Cut the small craft stick down to about 2″, drill a small hole 1¼″ from the end of the stick, and then glue the servo horn to the stick with hot glue.

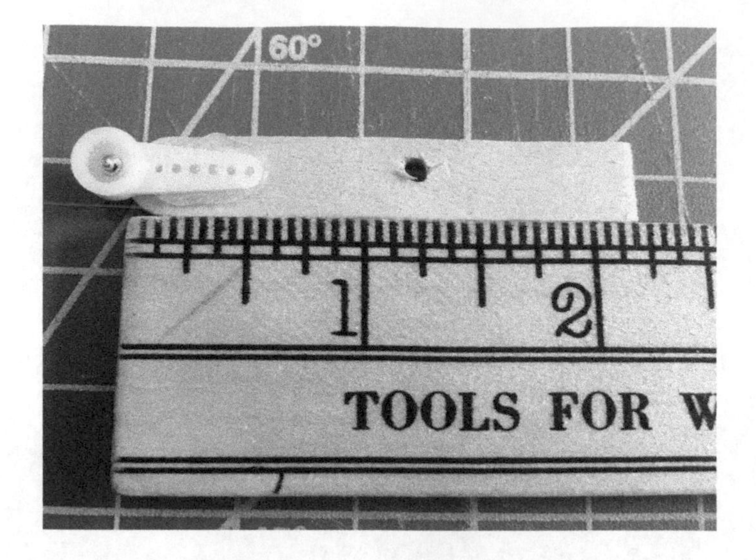

5. Press the servo arm and stick extension down onto the servo top, and turn it all the way to the left, with the servo oriented as shown in the following image. Reposition the stick extension (also shown here), then screw the servo arm into place using the short self-tapping screw provided with the servo.

6. Connect the servo to the RJ25 adapter. Refer to the earlier project in this chapter, "Head Turning Randomly Using 9g Servo and RJ25 Adapter," for details. Next, plug the servo into slot 1 and then mCore port 1.

7. Program and run the following.

The servo should rotate between the two positions.

8. Using needle-nose pliers bend the jumbo paper clip into the shape shown in the following image, with 2½″ legs on both sides.

9. Measure 1″ from the front of the box, and mount the servo to the inside of the box. Hot-glue it directly to the cardboard to make sure it's flat, as shown in the following image.

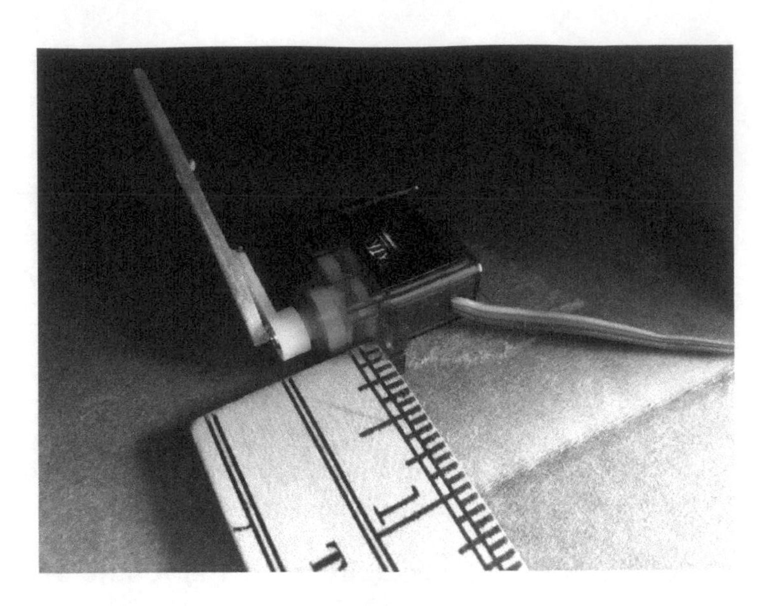

10. Hook the paper clip through the hole in the wooden craft stick, then rotate the servo toward the front of the box. Then tape the paper clip down to the flap.

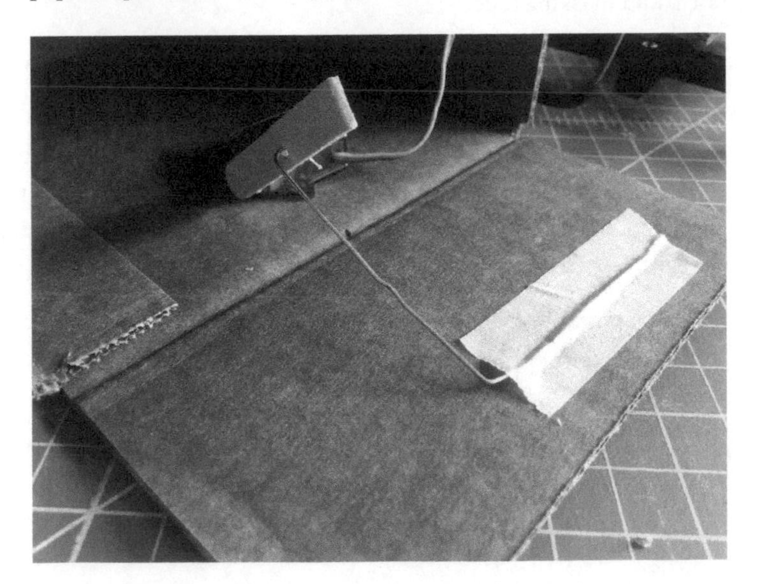

11. Your mouth should now open and close using your space bar as the trigger.

12. Once you know your mouth is working correctly, you can remove the tape and replace with a generous amount of hot glue.

Project: Rotating Eyes Using a 9g Servo and an RJ25 Adapter

For this project, you'll be using the same box from the previous moving mouth project along with a cardboard toilet paper tube, servo horn, and masking tape.

1. Take the tape with the drawn-on eyes off the flap of the box you built in the previous project.
2. Flip the box over, and then cut out the bulk of the flap with a hobby knife, leaving a ½″ border on three sides of one flap, as shown in the following image. This is where you'll be building your rotating eyes. Save the scrap, because we'll be using it in a future step.

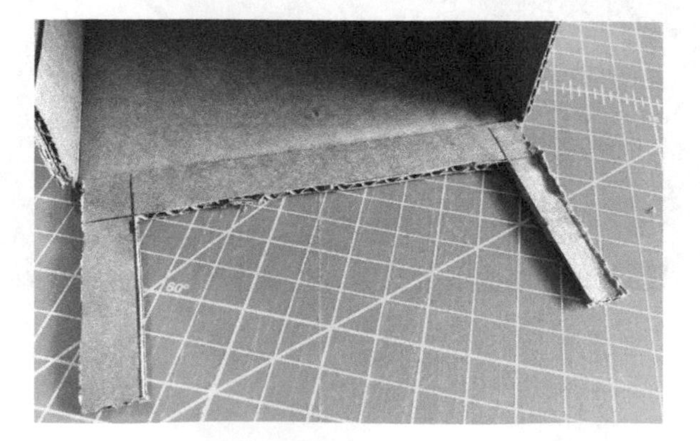

3. Take a cardboard tube, trace a circle around it on another piece of scrap cardboard, and then cut out that circle.

4. Hot-glue the servo horn to one side of the cardboard circle, with the part of the servo horn that attaches to the servo as close to the middle as possible.

5. Now hot-glue the circle onto one end of the cardboard tube.

6. Measure and make a mark on the upper-inside-left of the box 2¼″ down from the top and 1¼″ in from the front. Using a generous amount, put glue on the bottom of the servo and then glue the servo to the side of the box, lining up the bottom-right side of the servo with your mark.

7. Now, flip your box over and tape down the flap opposite the moving mouth on each side, as shown here.

8. On one of the scrap pieces of cardboard, measure 1 ½″ down on both sides and draw a connecting line. Center two quarters on the line, draw a line around them, and then cut them out. This is where the rotating eyes will line up.

9. Roll up some tape, as shown in the following image, and attach to the box over the rotating tube.

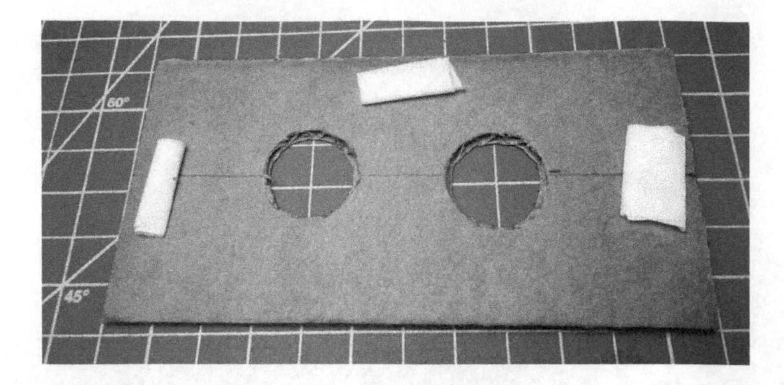

10. Write the code on Scratch and then run the program. The eyes will rotate between three positions.

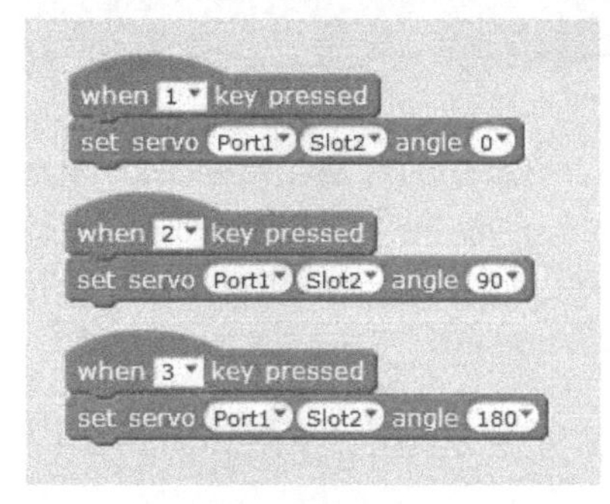

11. Using a pencil, draw the numbers 1, 2, and 3 on the tube through the openings you cut for the eyes that correspond with the three positions. You may need someone to press the 1, 2, and 3 key on your keyboard while you're numbering the three positions.

12. Remove the cardboard with the eye holes, and draw in three different eye shapes with a Sharpie. Again, you may need to have someone hold the 1, 2, and 3 keys on your keyboard while you're drawing in the eyes with the Sharpie.

    We also added a serrated edge to both sides to look like a mouth. Have fun customizing your own!

13. In Scratch, you can combine the rotating eyes and mouth movement using the following code.

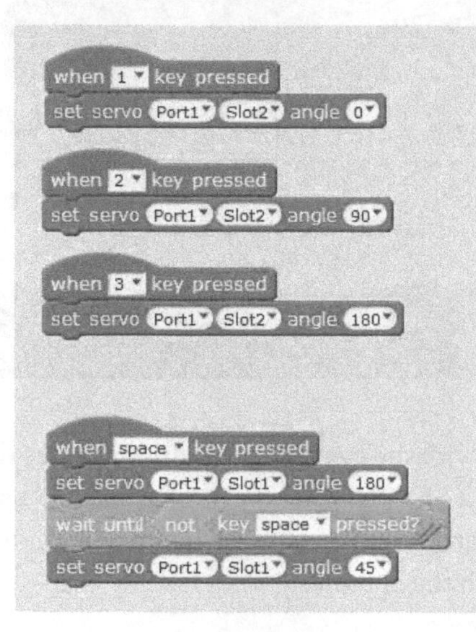

By pressing the 1, 2, and 3 keys, you'll move the eyes into the different positions. The space bar will open and close the mouth.

PUPPET MOVEMENT WITH SENSORS

The projects so far in this chapter have been preprogrammed for random or set movements without sensors. Now we're going to add some interactivity where your creature senses the environment and responds according to your program.

Project: "Feeding" Your Creature Using the Light Sensor

With this project, we'll use a light sensor that senses when your creature is "fed" and triggers a couple of motors to spin your creature's ears.

For this project you'll need two geared motors, a light sensor, and an LED. We'll attach the two motors inside the cardboard box, attach wheels on the outside, and then add some whimsical ears to the wheels. Then we'll affix the light sensor and LED inside the "mouth." When we "feed" the creature a piece of cardboard food, the ears will spin!

1. Starting with a fresh 6″ × 6″ × 6″ box or similar, tape the back of the box closed, and then cut off the side flaps on the other side, as shown.

2. Cut a ¾″ × 2″ hole in the bottom flap.

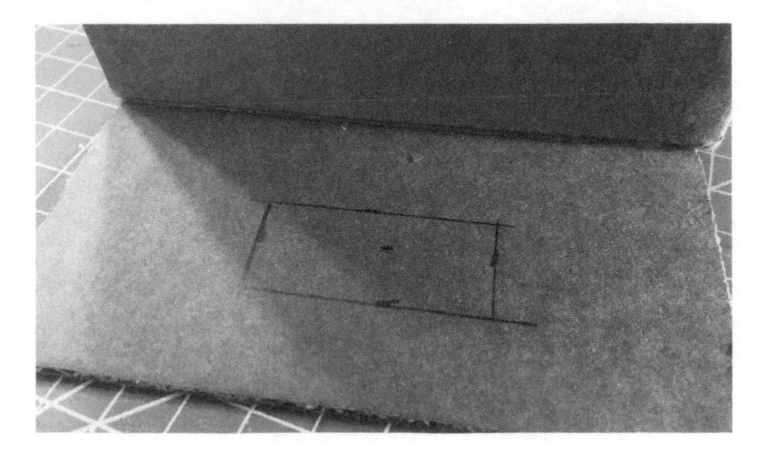

3. With a 1½″ hole saw, cut a hole in the upper corner of each side, as shown. I measured 2″ from the top and 2″ from the side.

4. If the hole looks messy, you can clean it up with a hobby knife.

5. It's easier if you mount the motors to a laser cut motor mount first. Laser-cut a mount out of acrylic using the template files available at the book's website. The additional holes in the mount are sized so that you can connect the motors to LEGOs or other Makeblock accessories.

> **NOTE** If you don't have a laser cutter, you can cut the mount by hand using a material of your choice, such as cardboard, thin wood, or thin sheets of plastic. PDF templates are available from the book's website at *www.airrocketworks.com/instructions/make-mBots*.

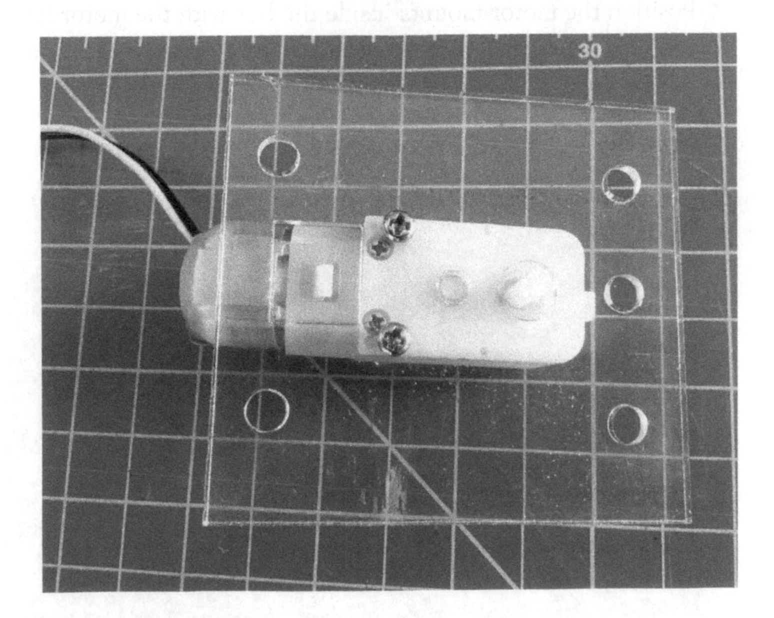

6. Put masking tape on the acrylic motor mount so that it's easy to remove later, and then add hot glue on top of the masking tape.

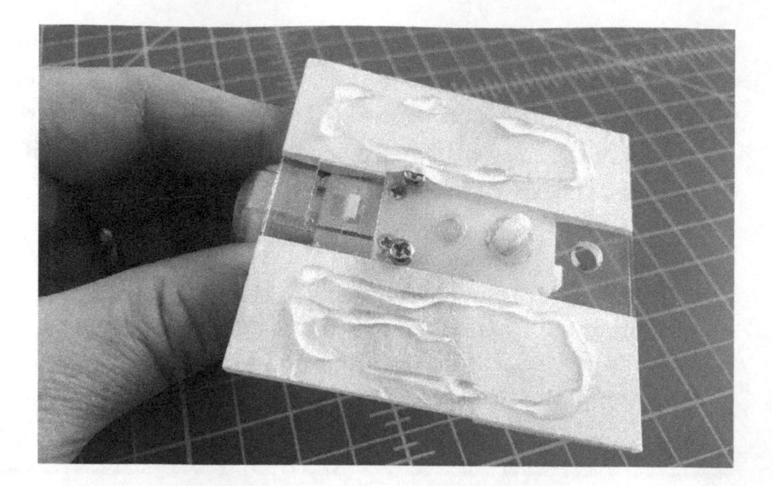

7. Position the motor mounts inside the box with the motor hubs centered inside the holes. We'll add the white plastic wheel hubs later.

   Now repeat steps 5–7 with a second motor.

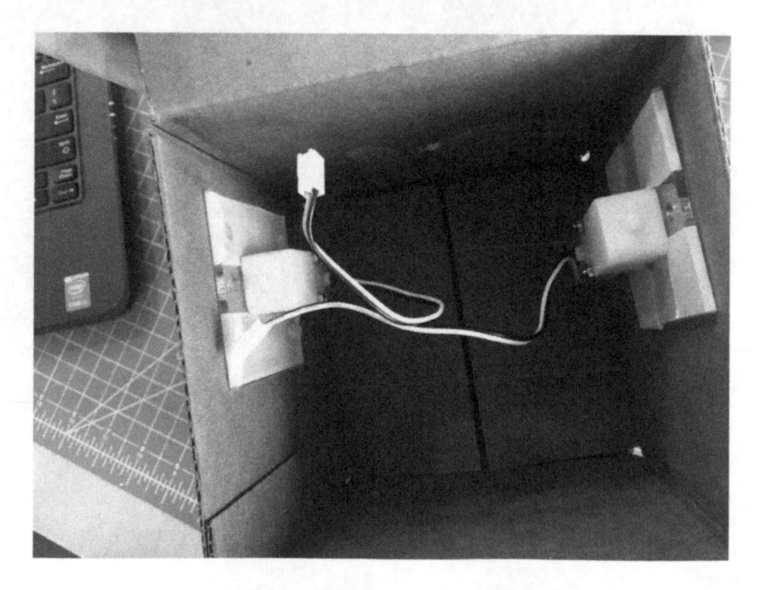

8. The moving ears that you'll be creating will spin around when the light between the RGB LED sensor and the Light sensor is blocked by a small piece of cardboard that is the creature's "food."

Once you have the LED sensor and Light sensor positioned, you can tape them in place. Then connect the RGB LED to port 1 of the mCore and the Light sensor to port 3 of the mCore. The motors can be connected to M1 and M2.

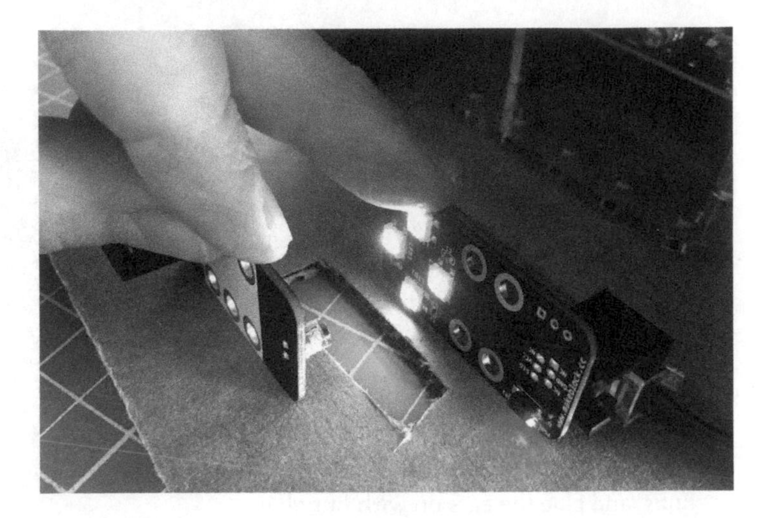

9. Code the following in mBlock and send to your mCore. The code will turn on the LED, then trigger the motors (M1 and M2) when the light going to the light sensor is interrupted.

```
when clicked
set led Port1 all red 20 green 20 blue 20
forever
  repeat 3
    if light sensor Port3 < 800 then
      repeat 3
        set motor M1 speed 255
        set motor M2 speed 255
        wait 1 secs
    else
      set motor M1 speed 0
      set motor M2 speed 0
```

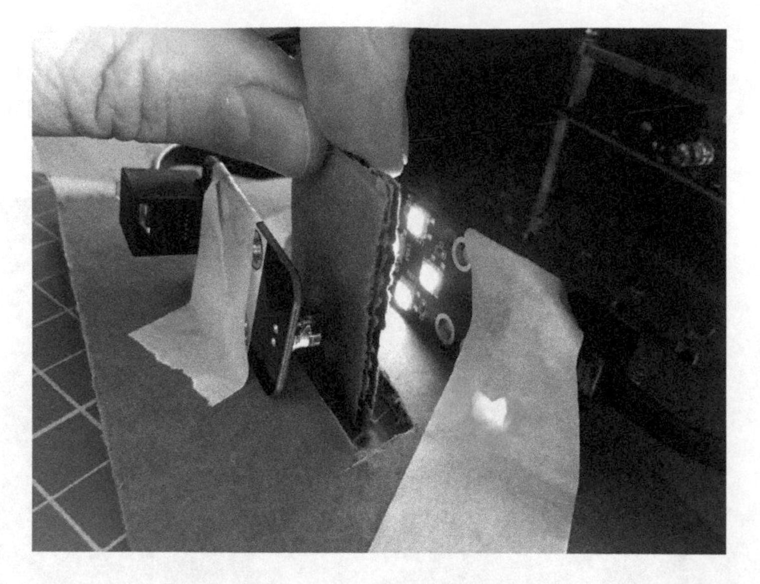

10. Now we'll add some fun ears to the wheel hubs using foam sheets. Cut the ears out of foam, apply masking tape to the plastic wheel hubs, and glue the ears on with hot glue.

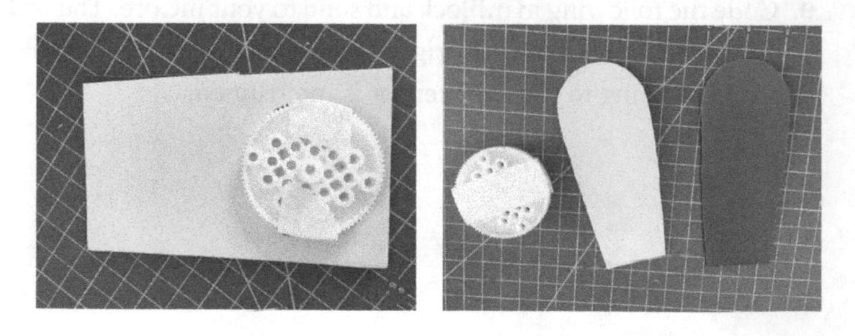

Now when you "feed" your creature by passing a cardboard disk through the mouth, the ears will spin. The creature will look like the one in the following picture. The creature also has an Ultrasonic sensor mounted on the front that will be part of the next project.

Project: Propeller Spins with Ultrasonic Sensor

For this project, you'll need the Ultrasonic sensor and one of the geared motors. When something or someone approaches your creature, a propeller on its head starts to spin!

1. Tape one side of a box closed. Then, drill a 5⁄16″ hole in the top center of a box.

2. Mount the geared motor to a laser-cut, acrylic motor mount and cover the motor mount with masking tape. If you don't have a laser cutter, you can cut the motor mount by hand using the full-size PDF template. You can make the mount from wood, cardboard, or soft plastic sheets.

3. Print the dowel-to-gear-hub adapter on a 3D printer using the template, available on the book's website: *www.airrocketworks.com/instructions/make-mBots*. Insert a ¼″ dowel connector into one end of the 3D-printed adapter and the other end onto the gear hub of the motor.

4. Add hot glue to the tape on the motor mount and glue to the top of the inside of your box so the 3D-printed adapter sticks out of the top.

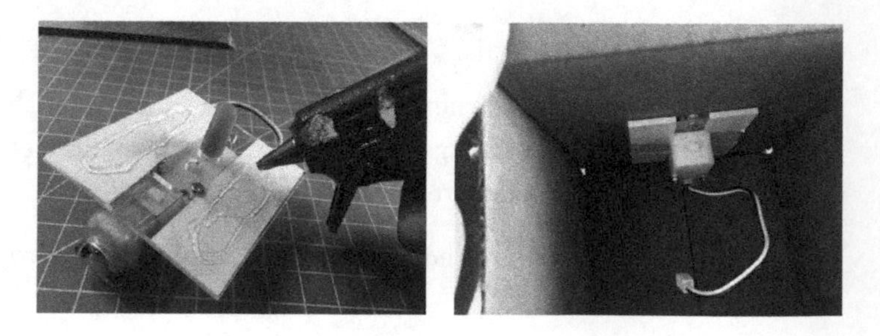

5. Cut a ¾″ × 2″ hole in the front of the upper front flap.

6. Cover the Ultrasonic sensor with tape, apply hot glue, and attach to the inside of the flap. The tape will protect the sensor from the hot glue and allow you to easily remove it later.

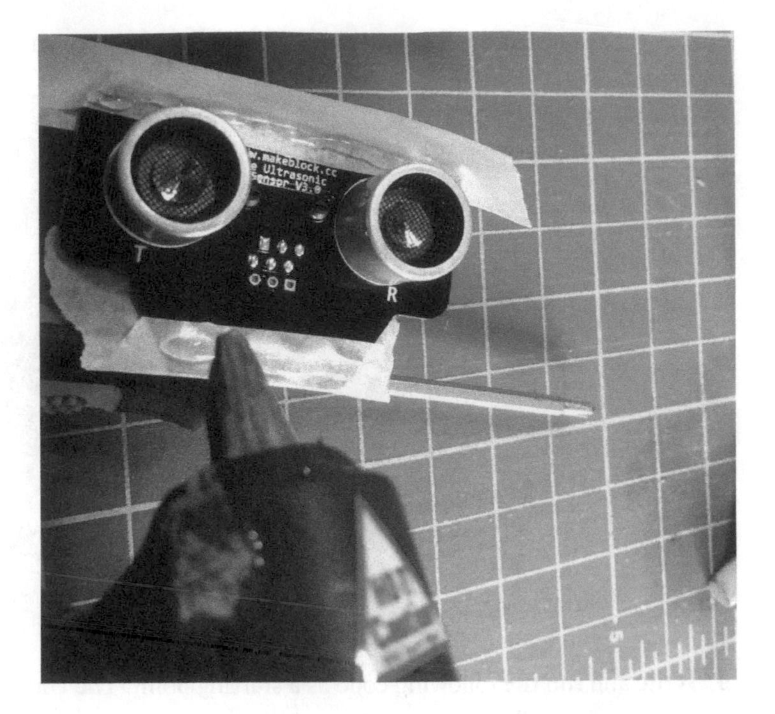

7. Mount the Ultrasonic sensor to the front of your box so that the "eyes" are exposed.

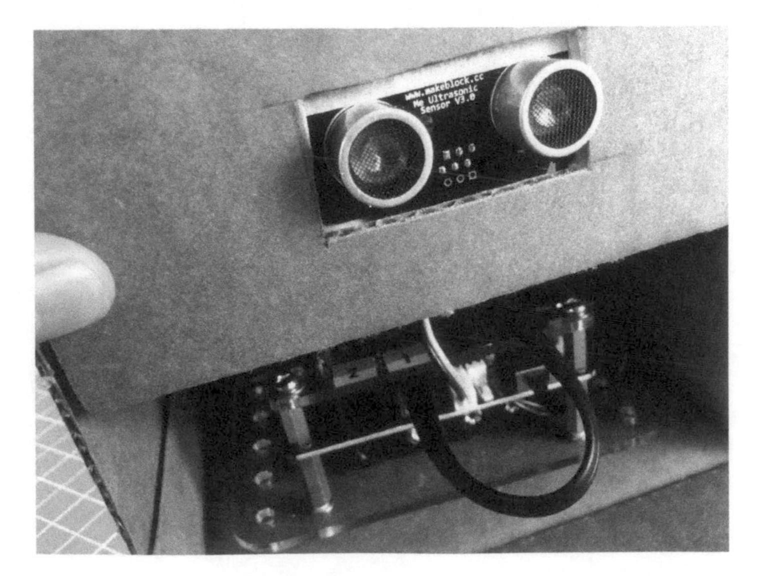

8. Connect the Ultrasonic sensor to port 1 and the motor to M1.

9. Write and run the following code as a starting point. The Ultrasonic sensor will sense your movement and trigger the motor. You can adjust the distance at which the Ultrasonic sensor begins to react. Here it's set at 20.

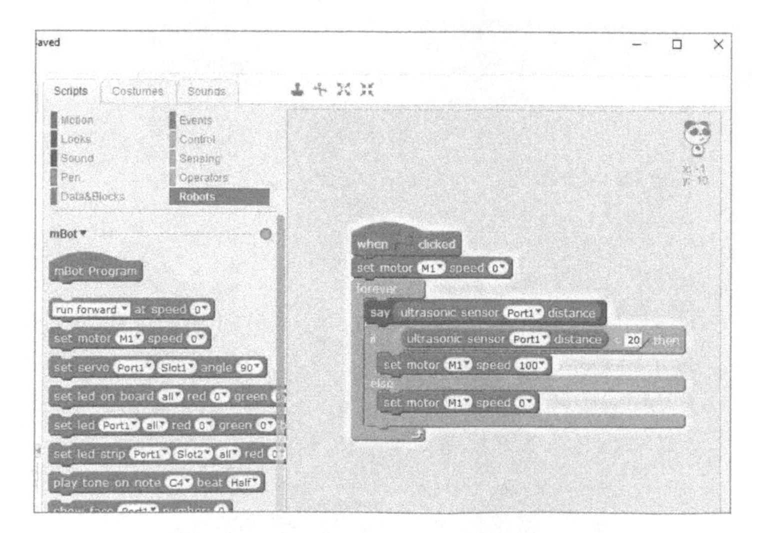

10. If you add a Say block in Scratch (in the Looks script) to your Ultrasonic sensor, you'll know it's working when the data input changes in the Panda's speech bubble. This one is set to 400. As you approach your creature, the motor should turn on.

11. I added a 6″ long, ¼″ dowel to the 3D-printed adapter and attached a propeller printed on card stock to the top. This is where you can have fun customizing your creature with whatever your imagination can come up with!

Project: Servo Arm with Paw Reaches Out When Motion Sensor Is Triggered

When someone approaches your creature, an arm linked to a servo will reach out.

Sometimes the circular motion of a servo or motor isn't exactly what you need for your creation. That's where mechanisms come in! Hundreds of websites exist to show you how to turn a simple circular motion into other motions. While there are quite a few options, the one we're going to focus on here is using a scissor linkage for a hand or paw that reaches out.

1. Grab eight large craft sticks and some brads to build the grabber.
2. Drill 5⁄32″ holes in the ends and middle of each crafts stick. It works well if you stack the sticks and drill them all together so you get the holes in the same place. Then insert brads, as shown. There should be one center hole with no brad.

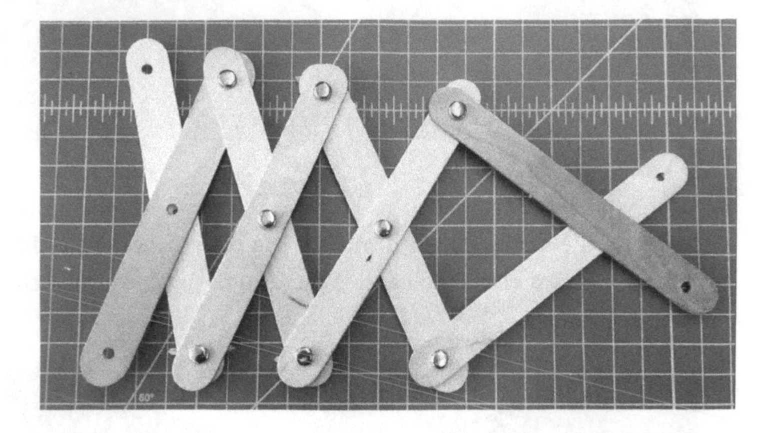

3. Decide which side of the box you want to attach the linkage to, or maybe you want two servos on each side! Cut a 3⁄4″ × 1 3⁄4″ slot in the box near the top center, as shown, then glue your servo to a laser-cut acrylic servo mount.

4. Cover the servo mount in masking tape, and then add hot glue and tape inside the box with the servo centered.

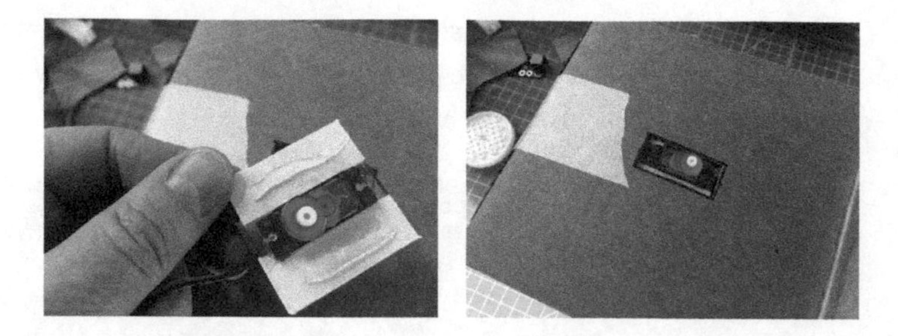

5. Grab your servo linkage arm. Flip the scissor mechanism over and glue the servo linkage arm over the third center hole that does not have a brad. You can see pictures of linkage arm in the following image.

6. Test-fit the servo horn onto the servo, then on the bottom end of the same craft stick as the servo arm, stack up several layers of scrap cardboard until the stick is about level with the servo, as

shown in the following image. Mark the cardboard through the hole so you know where to place a brad.

7. Cut down through the layers of cardboard with your hobby knife.

8. Connect the whole assembly using a brad.

This is how it should look from the side.

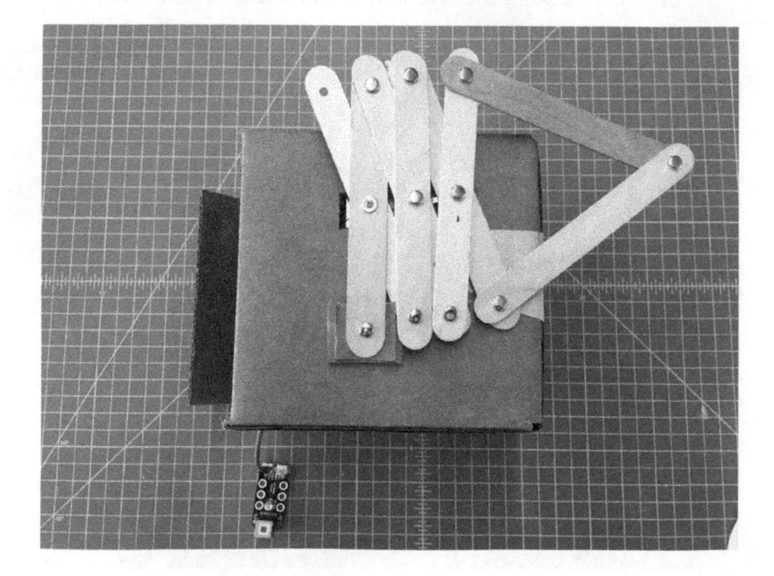

9. Using Scratch, create the following program, which will allow the Motion sensor to trigger the servo. Push your Motion sensor through a hole in the front of the box and connect the servo to the RJ25 adapter. Next, connect the Motion sensor to port 4 and the RJ25 adapter to port 1 on the mCore.

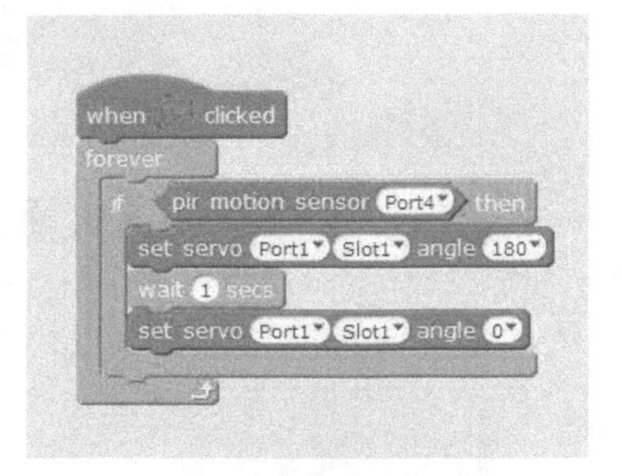

10. Once you have the mCore programmed and the servo lined up, you can permanently attach the servo arm using a screw and washer.

11. Now attach your claw, hand, or paw to the end of the scissor linkage and you're ready to go!

You can find some additional linkage and mechanism resources from the brilliant folk at the Tinkering Studio at *http://tinkering.exploratorium.edu/cardboard-automata*.

Project: Touch Sensor Triggers Scrolling Message

In this project, you'll make a message display when your creature is "petted"! We'll use the Touch sensor and an 8 × 16 LED Matrix display to make this happen.

1. Gather the components shown in the following image.

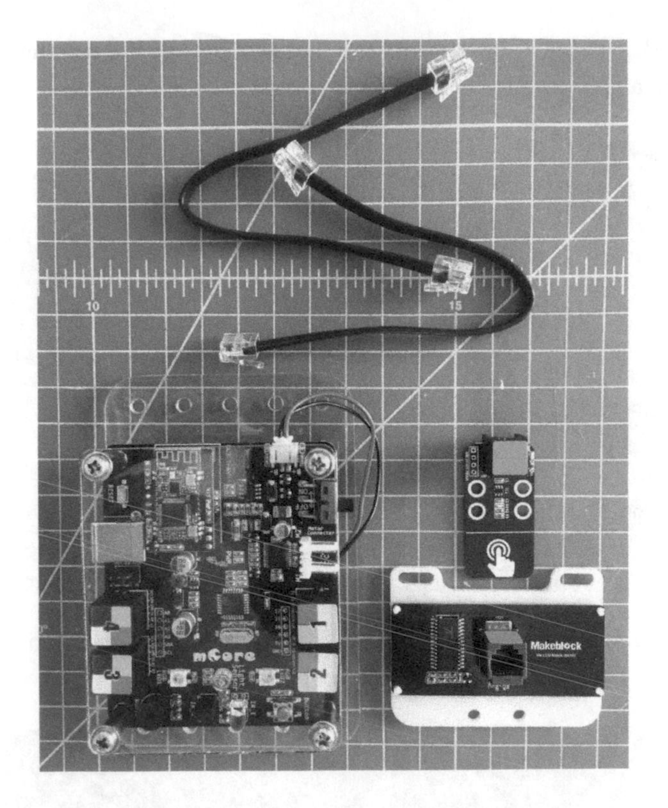

2. Grab a box, lay it flat, and cut off the two flaps opposite each other.

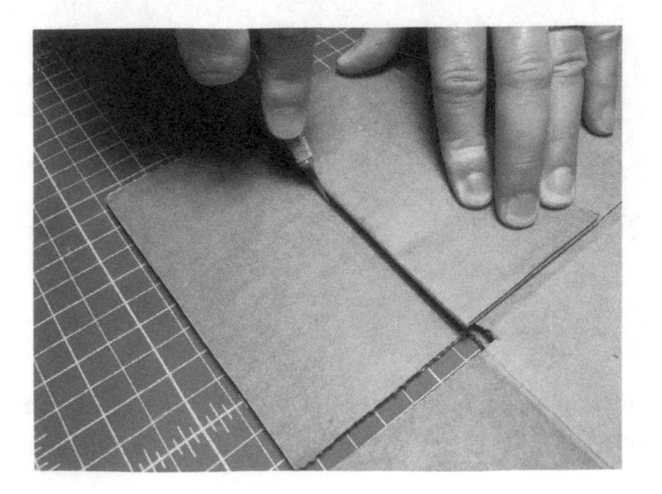

3. Mark a 1½″ × 1″ hole in the top flap and cut out the rectangle with a hobby knife. The LED Matrix will fit here.

4. In the middle of the bottom flap, cut a 1″ wide slit with a hobby knife. The Touch sensor will slip in here.

5. Add masking tape and then hot glue to the back of the LED Matrix.

6. Plug the LED Matrix into port 2 on the mCore and the Touch sensor into port 1.

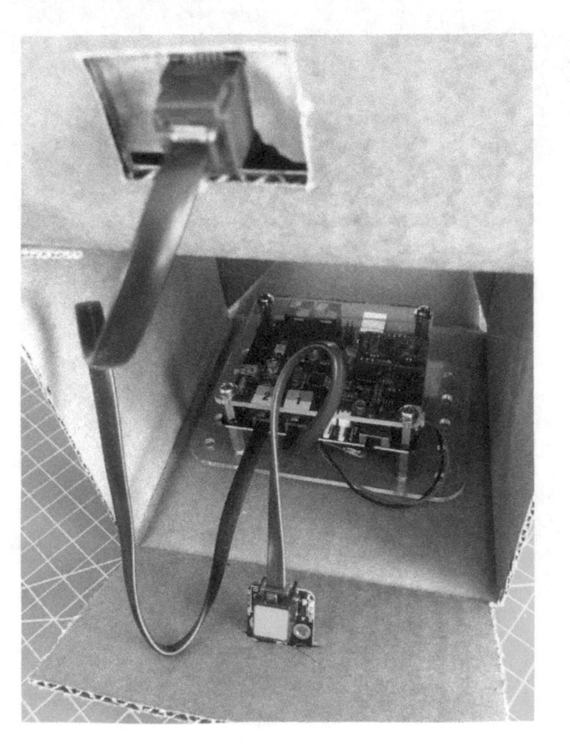

From the front it should look like the following.

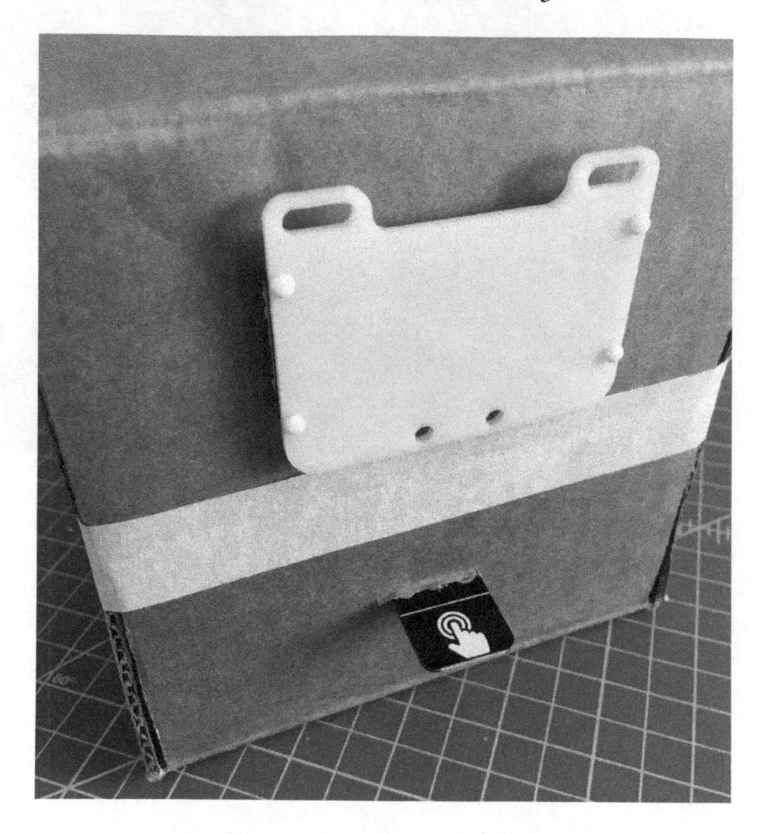

7. Write the following code in mBlock.

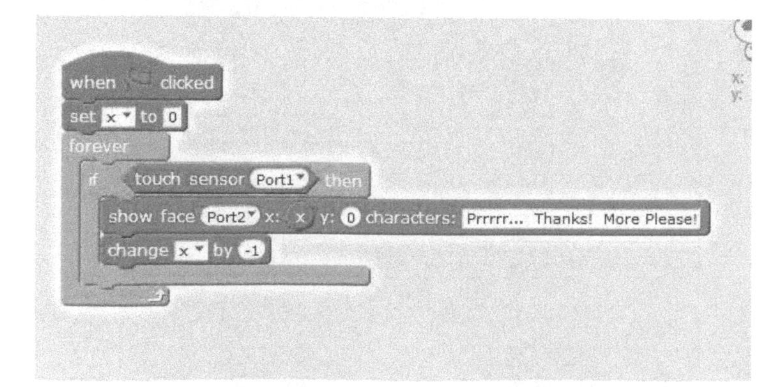

At this point, loading the program onto your mCore requires a bit more of an explanation (see the section in Chapter 1 titled, "Updating the mBot"). If you turn on your mCore and you hear three tones, you have the default program loaded. The default program includes all the files for your IR remote, line-following, and Ultrasonic sensor programs. These take up a lot of space and **do not** include the code needed to run the Touch sensor. You need to connect your mCore to your computer using a USB cable, open mBlock, and connect using whatever com port is available by going to the Connect menu, and selecting Serial Ports. Once you're connected, select Upgrade Firmware on the Connect menu and it should go through the upload process to load the software needed for all the sensors, including the Touch sensor. Now, when you boot up your mCore you should just hear just one short beep.

After upgrading the firmware and rebooting your mCore, connect your mCore via 2.4G serial or Bluetooth. Now, when the user triggers the Touch sensor by petting your creature, a message will appear on the LED Matrix! Each time you touch the Touch sensor, the message will start and stop. You have to keep touching the pad to see what the entire message says.

The projects in this chapter are just a starting point. Once you see how fun it is to set up and program sensors that trigger motors, servos, and digital readouts, you're limited only by your imagination. I've used a 6″ × 6″ × 6″ box for most of the projects in this chapter, but you can use whatever you have available, or what works for your particular project. One thing I've discovered: when you give kids lots of creative supplies like colorful foam sheets, cardboard tubes, boxes of various sizes, feathers, pipe cleaners, wood sticks, and other craft supplies, their minds and creativity come up with incredible things. The mCore and sensors provide the foundation for adding interactivity to any creative endeavor.

Measurement Devices

Computers count really fast. With small embedded computing systems like the mBots or an Arduino, anyone can create tools to record data about our physical environment. These measurement robots will work tirelessly for days at a time, and the data they provide about the world can offer young people a way to broaden their notions about the observable universe. The mBot is too large to be worn around a person's wrist, but the process for designing an environmental sensor for the mBot has much in common with developing technology that goes into wearable commercial tech products like smartwatches.

Probes and sensors have been a standard part of science classrooms for decades. Vernier manufactures dozens of different sensors designed to monitor everything from pH to turbidity. For more exotic measurements, like the composition of gases or liquids, specialized sensors are available, and are wonderful tools.

There are many specialized sensors available for the Arduino platform, and most of them can be used with the mBot. The techniques in this chapter can be generalized for most analog and digital Arduino sensors, but they are time consuming.

Here we've attached a Grove soil moisture sensor to the mBot using the RJ25 adapter board (see Figure 4-1). Many analog sensors operate with the same standard three wires: one for 5 V, one for ground, and one for sensor data. Grove's soil moisture sensor uses Grove's standard four-wire cable. Using any third-party sensor of this type requires matching the order of those pins on the sensor to the order on the RJ25 board.

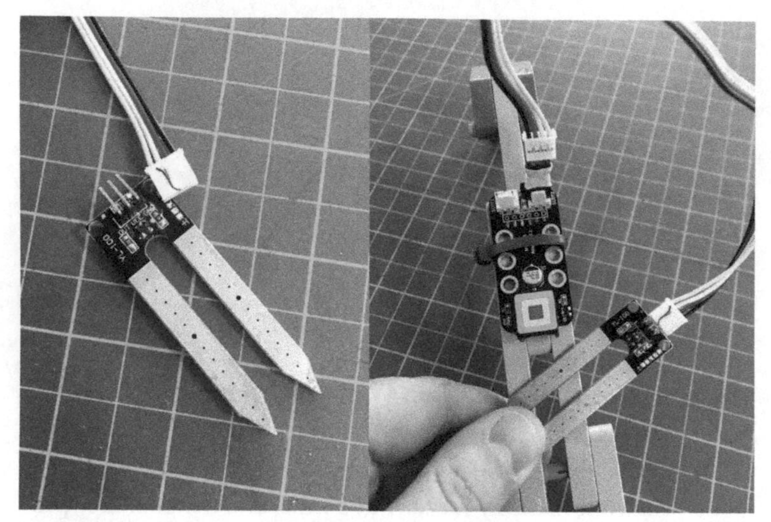

FIGURE 4-1: This soil moisture sensor from Grove uses a standardized 4-wire cable, even though the sensor only needs 3 pins.

With an easy-to-use generic tool like the mBot, the best learning experiences come from finding ways to use simple sensors to measure something specific and personally meaningful. Think of the long history of room alarm kits and toys sold since the 1970s. In every case, the core tech was something incredibly simple (a light sensor, a small button, or a magnetic reed switch) that became compelling when wrapped in the narrative context of adolescent spy fantasies. The goal of this chapter is to model how to think like both an engineer and a kid so you can construct the sensors you want out of the tools available.

In this example, we'll build a data-logging device that can operate independently for days using the mCore and a basic sensor, in both the Makeblock app and the mBlock programming environment. While these projects are built around a few common sensors, the techniques used to record, analyze, and export the data are consistent and reusable in most situations. The two examples we're building in this chapter were designed and built by elementary students who were studying how energy and resources were used in their school building. The flexibility of the mCore enables you to design and build devices that capture data to investigate your own super-specific questions, just like these kids.

The hallways into our building have two sets of doors, but some elementary students noticed that they were often both propped open. These students wanted to gather data on how often this happened, how long the doors stood open, and what effect that had on the hallway and classrooms.

This is close to the best case scenario for elementary students looking for local problems. Rather than focusing on a single question or measurement, these students had found a rich and complicated subject that could support multiple paths of inquiry.

Working together, they generated a large collection of measurement questions related to this drafty hallway (see Figure 4-2).

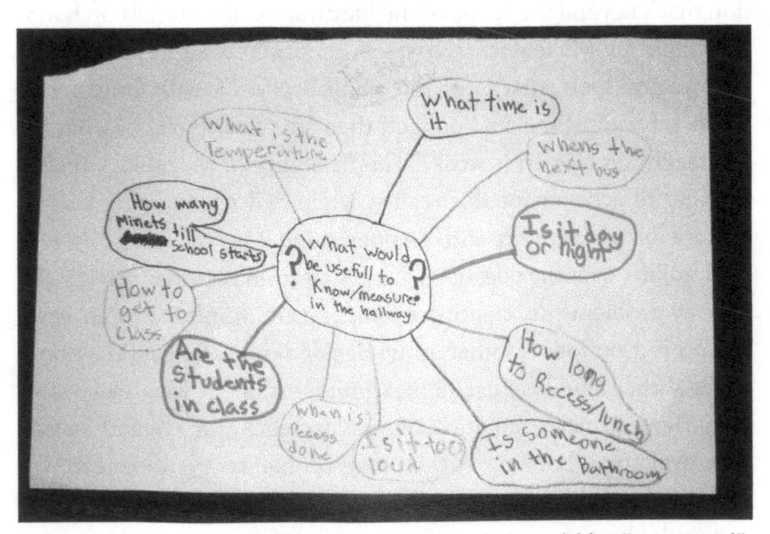

FIGURE 4-2: A student's brainstorm about what could be "measured" in the hallway, and how to use that data

It's easy to overlook this conceptualization step. It would certainly be easier and more direct to teach *about* a particular topic by gathering a supply of homogeneous parts and detailed instructions that walk everyone through building the same device. When I've had particular constraints on time or budget, I've chosen to run a single project with a class. However, I know that having everyone work on a single preplanned device causes the *excitement retention rate* (the number of kids who will still be enthusiastic and motivated by the end of the project cycle) to drop significantly.

Given the option, I will always choose to let students look for problems in the complexity of our daily lives, which we'll refer to as a *problem site* in our classroom. This is a much slower start than handing out premade kits, but it helps ensure that each kid starts wanting to know something specific and personal. What motivates kids out of the doldrums that invariably beset them in the middle of a project is personal investment in their unique questions.

The other risk of using open, student-directed inquiry with a physical computing project is that someone will latch onto a question that's beyond the scope of the hardware and materials on hand. Although the Makeblock ecosystem incorporates a huge range of sensors and tools, that's not the whole story. Given a finite set of sensors, boards, time, and budget, there are clear limits of what we can tackle "in class, this week." The flexibility of Makeblock makes it easier to account for this reality, while still encouraging kids to explore the problem site with an open mind.

Looking over the questions generated from consideration of the hallway problem site, creating a temperature monitor was an obvious project option. Another group decided to investigate the doors to see if they could get useful measurements about long they stood open. Both projects require some way to store and evaluate sensor data over time. Looking at these two problems will showcase two different methods for capturing and logging data using the mCore and mBlock.

Sensors report some specific bit of information about the world. The most basic program for any sensor is to display its output, which is recorded as part of a larger data set or used to trigger another action. But those tasks are almost impossible without a clear understanding of how the sensors report data and respond to changing conditions. Creating simple sensor display programs is a crucial step toward more elaborate projects and can help you create powerful stand-alone learning tools. This first section demonstrates how to display sensor readings using either a tablet or computer. The examples use Makeblock's thermometer, shown in Figure 4-3, but the principles apply to all supported sensors with numeric values.

The Makeblock app is a great tool for quickly creating control panels and status displays. Without access to Arduino libraries, tablet users are limited to the sensor blocks included in the Makeblock app. At the time of writing, this includes most of the sensors sold by Makeblock, but that may change as new sensors are released.

FIGURE 4-3: The Makeblock RJ25 connector board and water-resistant thermometer

One strength of the Makeblock app is the variety of tools provided to display sensor values. For these examples, we'll use a Numeric Display modeled on a common seven-segment LED, and the Line Graph. These are the best tools in the Makeblock app for providing precise history for a sensor.

Open the Makeblock app and create a new Makeblock sketch, then choose the Line Graph display block from the Custom palette (see Figure 4-4).

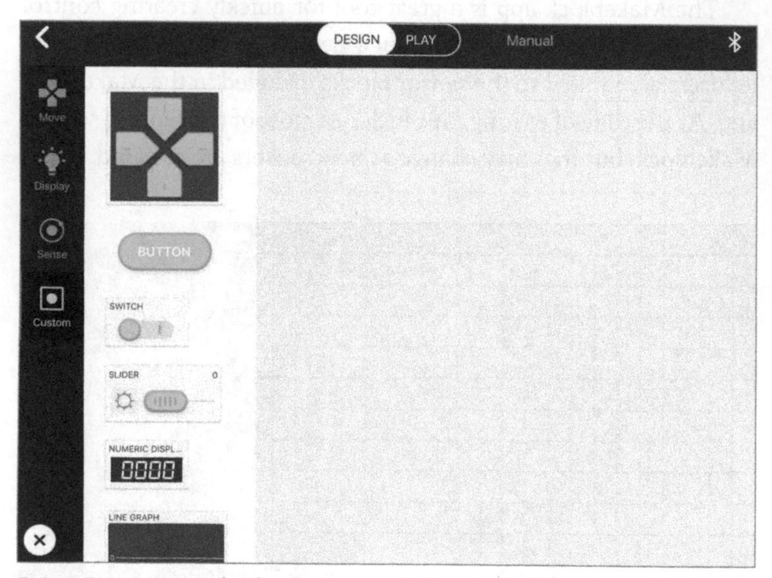

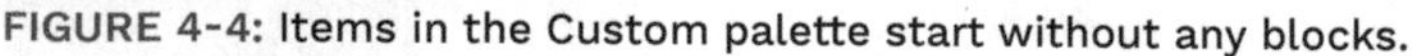
FIGURE 4-4: Items in the Custom palette start without any blocks.

At the top of the Makeblock app screen, there's a button that allows you to toggle between Design and Play mode. To add elements or make changes to them, you need to be in Design mode. Play mode will activate those elements, along with any scripts under a When Start hat. In Scratch parlance, blocks with the swoopy tops are called *hats*, because they must sit at the top of a stack of blocks. All hat blocks and the blocks attached to them require some signal to activate. Makeblock also has a variety of interactive UI elements that trigger specific behaviors when selected. Displaying sensor values requires a UI element such as the Line Graph, Analog Meter, and

Numeric Display, but you can use the Read Sensor block anywhere in a Makeblock app program.

For this example, we're using an mCore with the three-wire temperature sensor, connected through the RJ-25 adapter board, as shown in Figure 4-5. The temperature sensor is analog, so it needs to attach to port 3 or 4. In this example, we're using port 3 and slot 1.

FIGURE 4-5: The temperature probe connects to the RJ25 board, which bolts securely to the LEGO Technic frame from Chapter 1, "Kit to Classroom."

To model all of these connections inside the Makeblock app, we need to add code to our blank Line Graph block. In Design mode, select the Line Graph block and choose Code. (See Figure 4-6.)

The block interface for the Makeblock app offers many tools to work with, but this example requires only a few. For a more detailed discussion of how to program with the Makeblock app UI, revisit Chapter 2, "mBot Software and Sensors."

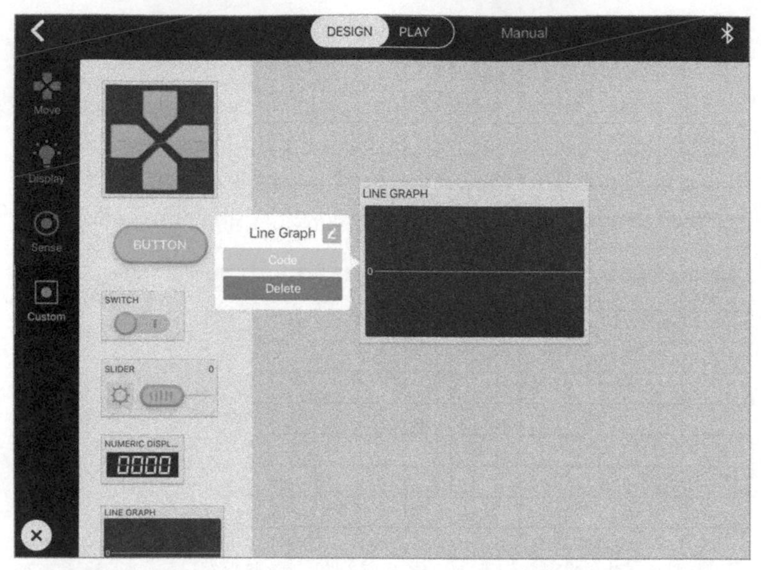

FIGURE 4-6: The Context menu that opens the block-based coding interface is available only in Design mode.

We'd like the line graph to constantly display the temperature reading. In the Detect palette, there's a Read Common Temperature Sensor block. Drag it out, then configure the Port and Slot values to port 3 and slot 1. Wrap that orange block inside a purple Display On This block from the Display palette. Place that purple block inside a pink Repeat Forever block from the Control palette. Finally, attach the Repeat Forever underneath the light blue When Start hat. The final block is a multicolored monster, as you can see in Figure 4-7, but it will grab and graph the thermometer data indefinitely.

Tap the back arrow in the top left of the screen to return to the UI design screen. At this point, it's worth renaming the Line Graph block to reflect what it's actually graphing. Tap the Line Graph block and use the orange pencil icon to rename it.

The line graph and all other display elements will update after the Makeblock app is in Play mode.

If the temperature is stable, the graph will show fluctuations within a few degrees Celsius. If you take the probe from your warm hand and plunge it into a cup of ice, the graphs will zoom out to

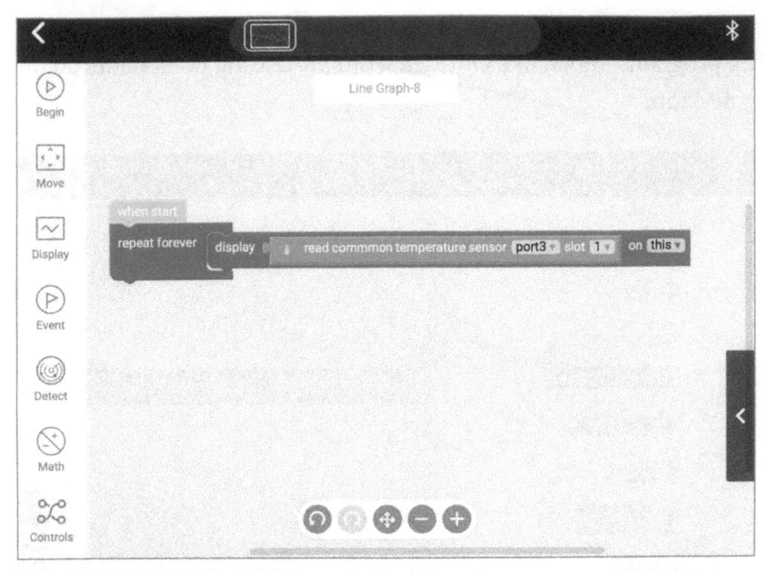

FIGURE 4-7: When creating custom elements, you can set sensor locations directly inside the blocks without using the Port interface.

display the sudden dip in temperature. Scaling keeps major temperature swings visible for a while, but the detailed values are obscured. The variables in the Makeblock app allow programs to track specific values, even after they scroll off the dynamic line graph. Next, we'll create a variable to capture and display the lowest temperature recorded by the sensor.

Blocks that relate to variables in the Makeblock app are Change Item By, Set Item To, and Item, and they all appear in the Math palette. (See Figure 4-8.)

Once any of those blocks are dragged out of the palette, clicking the pulldown after the word Item will show all existing variables and allow you to delete or rename them. For experienced programmers, it might seem a bit weird that there's no Create Variable option in this menu. In the Makeblock app, new variables are always called simply *Item*. After a variable is renamed to something helpful and clear, dragging out another block from the Math palette will create another bland Item. In fact, the Makeblock app doesn't offer any way to create a new variable other than renaming! This system attempts to avoid situations where many anonymous Item variables clutter up programs, much like Untitled word processing documents do on a desktop.

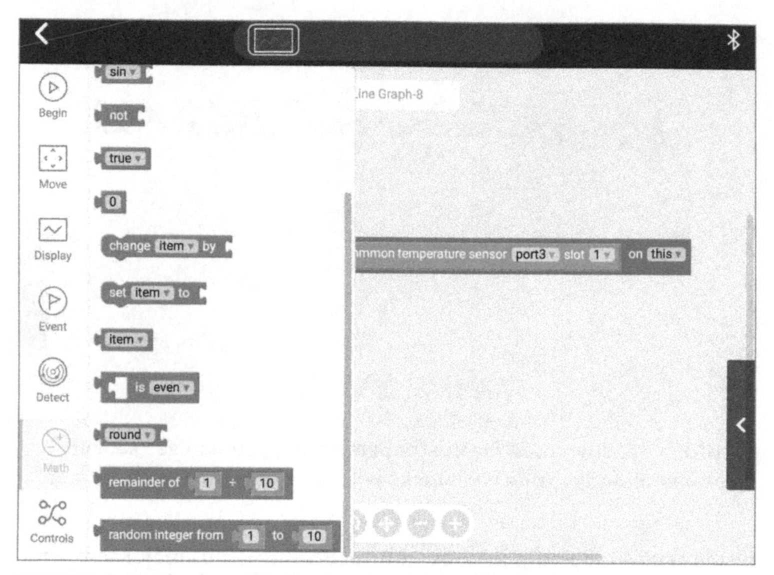

FIGURE 4-8: Blocks with the keyword Item live in the Math palette and offer access to all variables in the Makeblock app.

It's important to recognize that these variables are not tied to particular UI elements! The Makeblock app's system of bundling code to particular buttons or displays can obscure the fact that all the code is part of a single program and runs simultaneously. Variables can be set, read, and modified across the code blocks of several elements, which will help keep the Lowest Temp program visually cleaner.

One of the real hassles of block-based programming is screen width. When blocks nest inside each other, it's easy for important information to get pushed off the right side of the screen. Horizontal scrolling is a pain! To avoid this, it's good practice to use a variable to store the temperature reading, instead of calling the sensor several times in a single loop (see Figure 4-9).

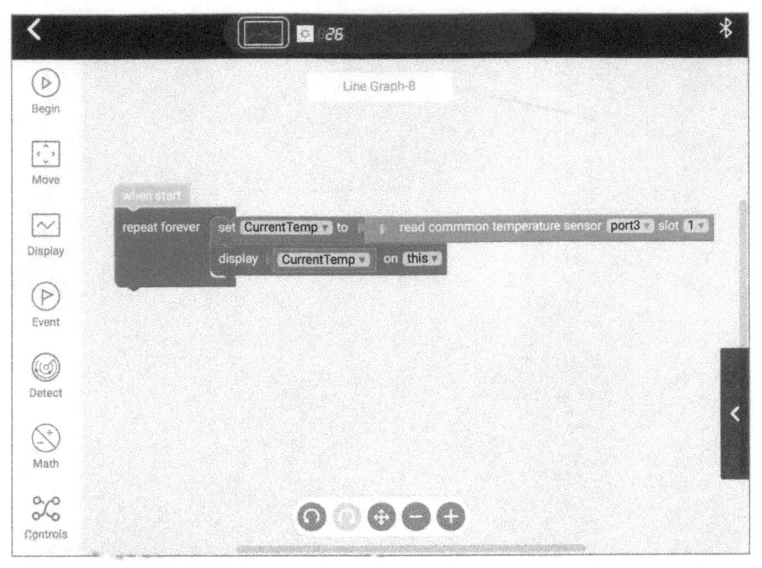

FIGURE 4-9: Saving the temperature reading in a variable makes it possible to build another UI with confidence that both will use the same data.

This technique transforms a long block like Read Common Temperature Sensor on port 3, slot 1, into one compact variable name CurrentTemp. Each time we use a variable instead of a sensor call, the program gets more consistent (because the values don't change each time), more responsive (because the program running on the tablet doesn't have to wait to hear from the mBot), and more clear.

Using the Makeblock app, we've created a portable thermometer that displays of-the-moment data on the screen of a mobile device. But if something interesting happens with the temperature while no one's watching the screen, there's no record of it. To fix this, we'll create a second variable to store the lowest temperature observed by the sensor. We'll also create a new UI element, Numeric Display, that will always show the lowest recorded temp from the current session. See Figure 4-10.

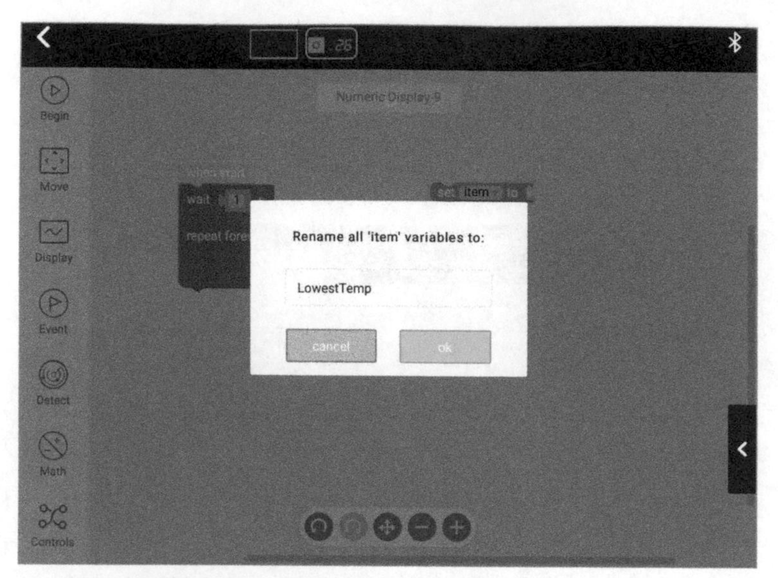

FIGURE 4-10: To create a new variable in the Makeblock app, drag any block that uses the keyword Item onto the stage, and then rename the variable.

Although the lowest temperature won't often change, it needs to be constantly checked against the CurrentTemp threshold (see Figure 4-11). Placing the Wait block before the comparison loop ensures that a valid CurrentTemp variable will always be available.

The blocks in this Numeric Display never set the value of CurrentTemp. The Line Graph block is still updating the CurrentTemp value during every loop, and this block can make use of that data. It's worth noting that there's no strict sequencing between the code in the Line Graph block and these blocks. Both use a RepeatForever block and will loop independently. Loose coupling like this might cause problems with data that changed rapidly and non-linearly, like the ambient noise level at a concert. Because we are working with something slow and steady like temperature change, though, there's little measurable difference if the LowestTemp loop checks the same CurrentTemp value twice, or if the CurrentTemp value updates quickly between loops of the LowestTemp check.

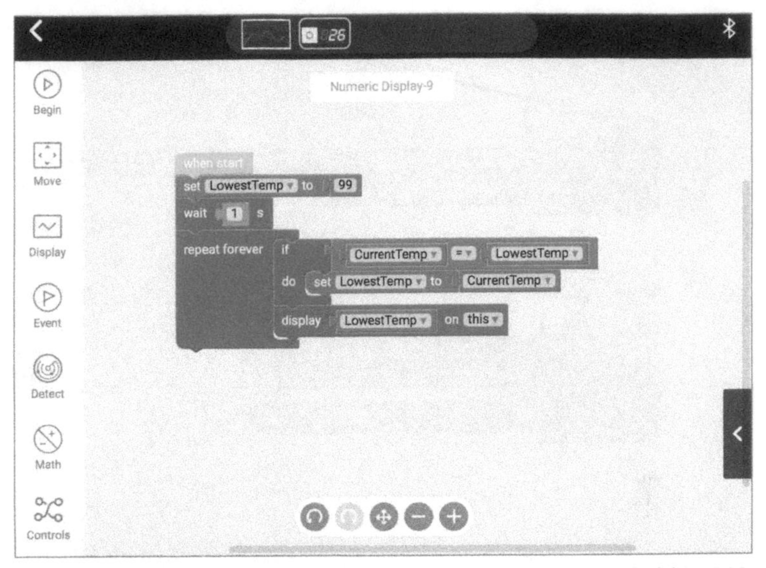

FIGURE 4-11: Why start off by setting the LowestTemp threshold to 99? Having a large initial value ensures that on the first loop, CurrentTemp will be below LowestTemp and the Do section will execute.

When introducing the Makeblock app, it's helpful to attach each script to the relevant UI element. If the widget isn't displaying the LowestTemp properly, checking the code inside the Numeric Display UI element is a good first step (see Figure 4-12). You can quickly move between the scripts attached to all elements in a program by using the horizontal navigation panel at the top of the screen. However, blocks from anywhere in a program can change or update each individual UI element. In more complex programs, it might be more elegant to keep all the scripts in a single location.

If you use the wide range of sensors available, the Makeblock app can become a powerful and versatile data-monitoring station. The only hard limits on the Makeblock app as a research tool are the range of the Bluetooth connection and insufficient, large-scale data storage. Using mBlock, the desktop programming tool, offers ways around those specific limitations but presents a very different experience when working with sensors. In the next section, we'll re-create a similar temperature-logging program within mBlock and highlight the data-related tools within the Scratch environment.

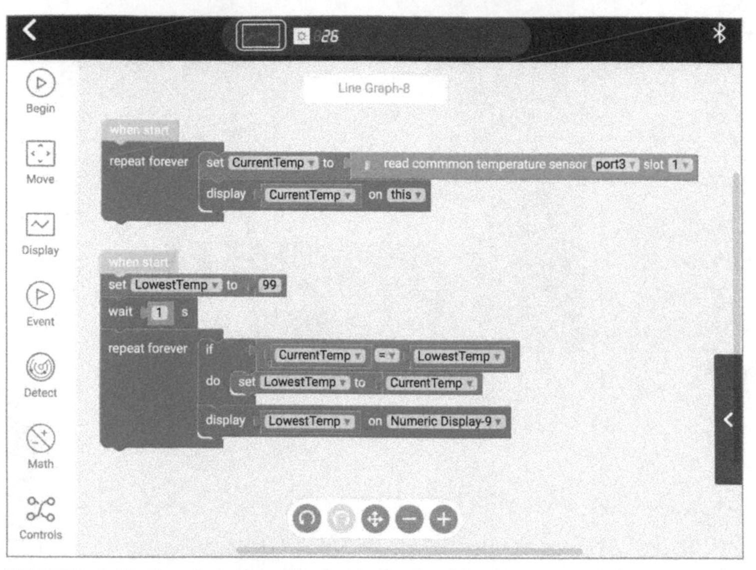

FIGURE 4-12: By changing the last display block to name the Numeric Display instead of the self-referential keyword *This*, the entire program can attach to one element.

MONITORING SENSORS IN MBLOCK

When you're using the mBlock programming tool on the desktop, the easiest way to check the value of a sensor is to use the purple Say blocks. It isn't fancy, but it is still a great way to check that the values match your expected sensor behavior. When working with a class or large group, we require a Say sensor test as the mandatory first step for any program. This little stack of blocks serves as a safeguard against common start-of-project problems. If the polar bear can say the temperature, like she's doing in Figure 4-13, then we know that the mBot board is powered properly, the serial connection works (whether it's Bluetooth, 2.4G, or USB), all the sensor wires are connected properly, and the displayed data matches expectations.

Creating this simple program replaces a lengthy and dull preflight checklist that leads to a properly configured mBlock and a blank screen. By using Say blocks to fix low-level issues, we also create an interactive tool that helps students experiment with the hardware.

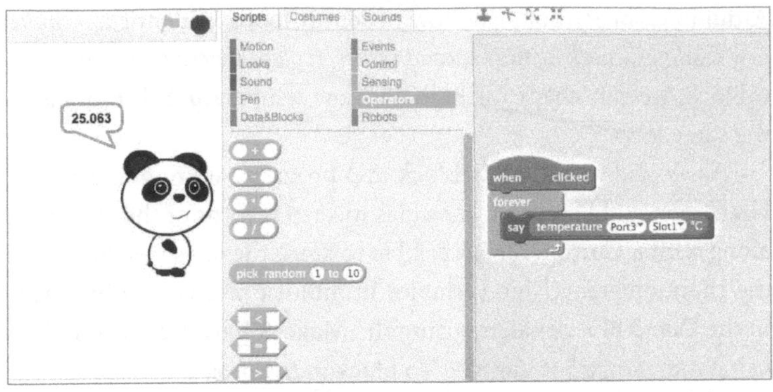

FIGURE 4-13: The default mBlock panda sprite doesn't do much in our programs except shout sensor values.

Since the mBot's wireless communication allows us to measure the temperature far away from the controlling laptop, it's important to display the temperature on the device itself, not just on the mBlock stage. We'll opt for the conventional 7-segment display. Just like the Say block, the 7-Segment block will accept any alphabetic or numeric value, including variables or mathematical expressions, as shown in Figure 4-14.

In mBlock, you can use variables to track record temperatures, although the visual syntax looks different from the Makeblock app. Moving between two different block-based tools teaches new programmers to look for structural similarities underneath syntactic differences. Whether you're working in the Makeblock app, mBlock,

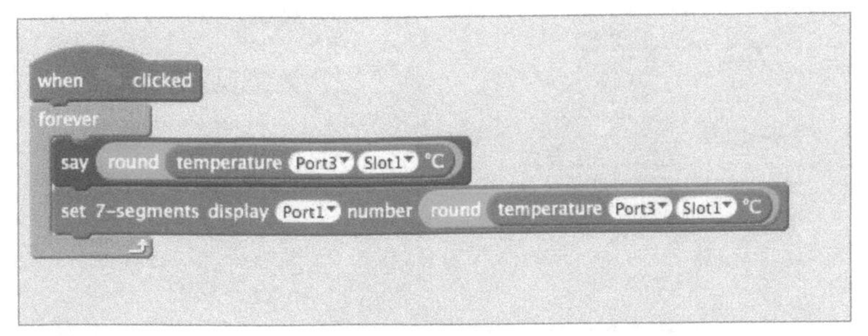

FIGURE 4-14: Use the green Round block from the Operators palette to round the temperature to whole degrees.

Arduino C, or Python, the flow of sensor loops is identical: *acquire new readings, check against special values, replace if necessary, repeat.* In mBlock, we can check for high and low temperature thresholds in the same loop.

As we did in the Makeblock app program, we're going to use HighTemp and LowTemp variables to keep records of the extremes, along with a CurrentTemp variable to store the latest reading from the thermometer. Using variables in mBlock means creating them in the Data&Blocks palette using the Make A Variable button, then using the orange Set Variable To block in the program itself.

This program starts off by setting both the high and low temperatures to the initial reading from the thermometer. This provides a reasonable baseline to check against over time. The default value in the mBlock Set Variable To block is 0, which isn't a neutral value for temperatures in Celsius, much less Fahrenheit!

Once the variables are set, we need to establish the main loop of the program. As in the Makeblock app program, this loop will run continuously once the program starts, and require no interaction from the user (see Figure 4-15).

I recommend that you use an If/Else block in your loop to check the first criterion, and then nest another If block in the Else statement. This ensures that no CurrentTemp value replaces both the

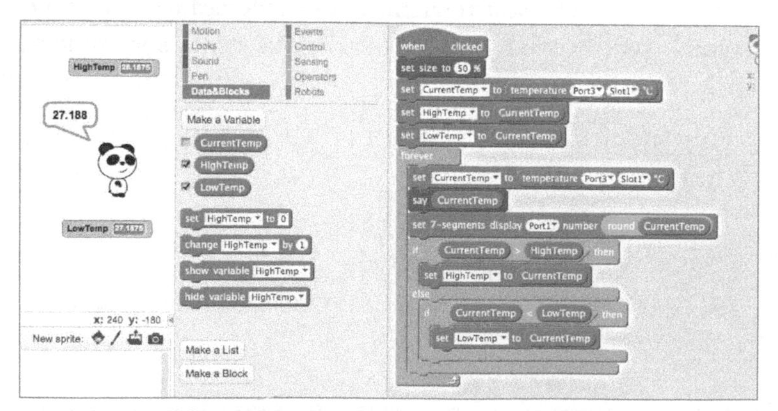

FIGURE 4-15: This version reproduces the behavior of the mobile program created with the Makeblock app.

HighTemp and LowTemp records in a single loop. When the program starts and HighTemp and LowTemp are assigned the same value, those values won't change until the temperature shifts.

At the beginning of each loop, the value for CurrentTemp is displayed on the physical 7-segment display and on the screen with the Say block. The on-screen displays for HighTemp and LowTemp are controlled by the small checkboxes next to the variable names in the Data&Blocks palette, or by using the orange ShowVariable and HideVariable blocks in a program.

One of the enduring charms of Scratch is that the general-purpose tools are strong and flexible enough to make up for the lack of built-in functions that provide instant answers to very specific challenges. There's no Line Graph function in mBlock that matches the one in the Makeblock app. Instead, there's the wide-open Pen tool that can create everything from mathematical drawings to 3D environments and everything in between. There's no end to the complexity and challenges that can emerge from trying to create graphical representations of data with the Pen tool. This simple line graph is an invitation to experiment and create with this incredible tool, shown in Figure 4-16.

The mBlock Stage area becomes the XY-coordinate grid and sets the position of the cursor based on the data from the temperature sensors (see Figure 4-17). The value of the horizontal X-coordinate needs to steadily increase so that the graph moves steadily to the

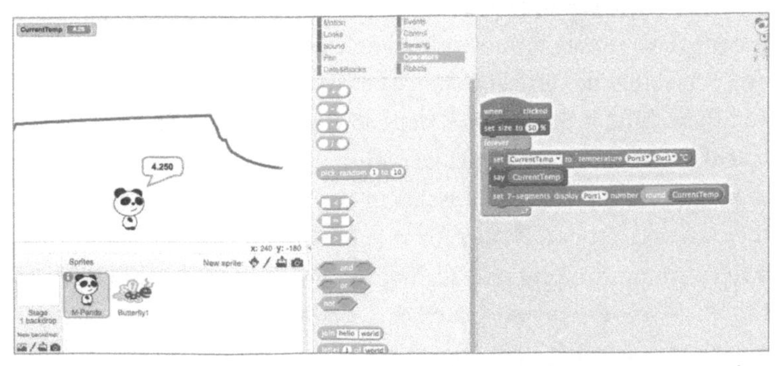

FIGURE 4-16: While the panda sprite is shouting the temperature, the small butterfly sprite is drawing the line graph.

right over time. This example uses a five-step "tick" for the graph, meaning that the distance between each data point is five steps on the Stage. Using fewer horizontal steps in each tick would create a slower and denser graph, whereas a larger tick would result in bigger movements on the graph over a smaller time interval.

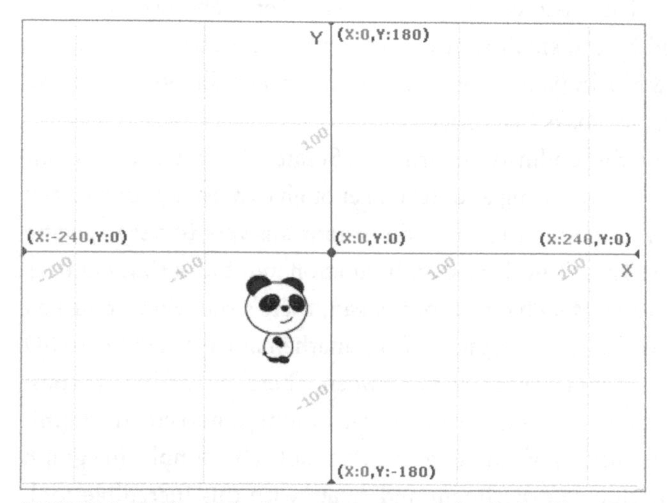

FIGURE 4-17: Scratch chose a center-original coordinate plane to better match the way graphing is taught in elementary math classes.

A normal temperature range of a few degrees doesn't create a very dramatic graph. In mBlock, the XY-coordinate grid has a center point of (0,0) and has a maximum Y-coordinate value of 180. We can adjust the temperature value to display better within that coordinate system. This example, shown in Figure 4-18, scales the temperature by 3.5, making the vertical distance between degrees more noticeable, and then shifts it down by 75 steps so that 0 degrees is in the lower half of the stage. These particular constants make sense for an outside thermometer reporting in Celsius in certain parts of the world, but they would be a bad choice for monitoring the temperature inside a refrigerator. Shifting and scaling the graph to make good use of the Stage real estate and match the conditions being measured is a meaningful design task in and of itself.

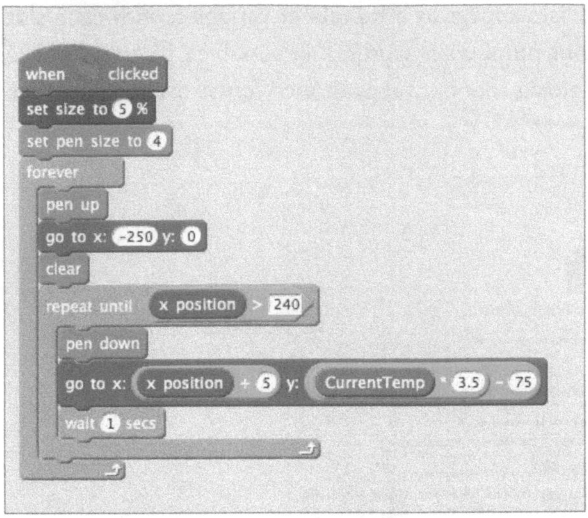

FIGURE 4-18: This program draws a line graph and resets when the sprite exits the right side of the Stage.

So far, we've been using the mCore to report on momentary observation. While we've used visual displays and record-keeping to broaden the moment of observation, the raw data is ephemeral.

Arduino Uno–derived boards like the mCore are memory-limited, making it impractical to use a plain board for long-term data monitoring. The many data-logging projects that use Arduino-derived boards must all record data to some form of external memory, either by a connection to a computer or the addition of SD cards. The specialized form factor of the mCore makes it more difficult to use one of the many Arduino shields that add SD card storage. But since the mBot can use the standard persistent Bluetooth or 2.4G serial connections to send back temperature readings to mBlock, the hard work of data storage is already done!

So far, the thermometer we've built hasn't needed to store or evaluate large amounts of data. It simply took the temperature reading from the sensor and sent it to the 7-segment display, and checked that value against dynamic thresholds. Tracking all of the temperatures over time means that we need to use mBlock's List data structure.

In mBlock, lists appear as a bundle of variables, with each slot capable of holding numbers or words. Like variables, List blocks have a special set of blocks that only appear once you've created one. (See Figure 4-19.)

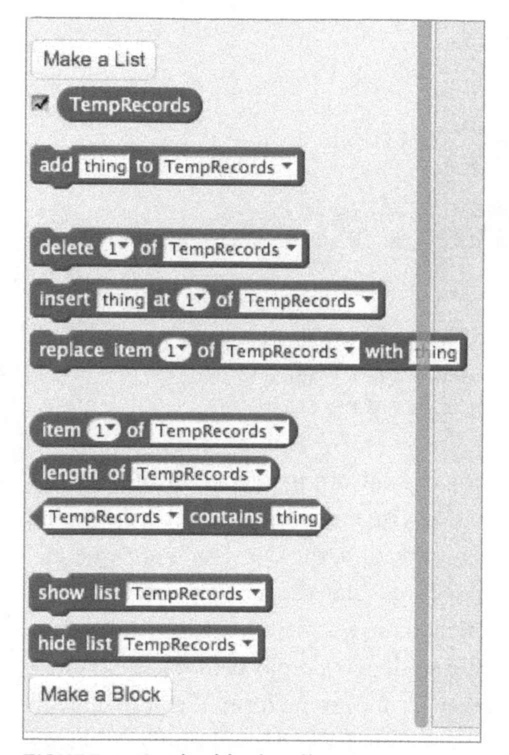

FIGURE 4-19: List blocks allow programs to read, insert, or remove individual items from any position in the list.

Lists start empty, but do not reset or clear unless you use the List block named Delete All (see Figure 4-20). This persistence can be a useful feature if you're trying to capture data over a long time, but when you're designing and testing it's helpful to clear the list at the beginning of the program. By default, mBlock displays Lists values on the Stage, but the actual data will quickly outstrip those tiny windows.

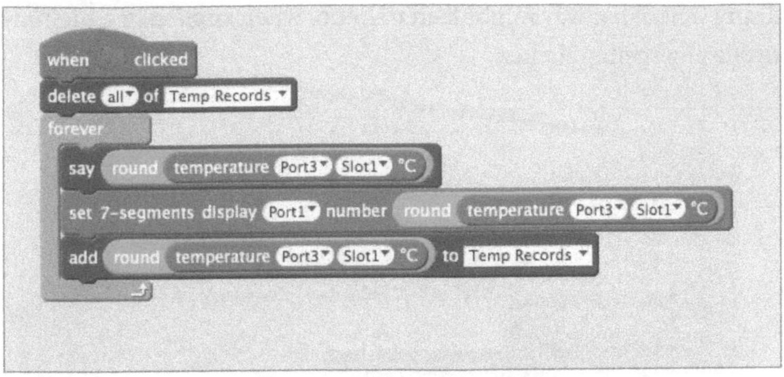

FIGURE 4-20: Without the opening Delete All block, new items would append to TempRecords each time the program ran.

After gathering a large data set, we can export the data and use other tools for analysis. This is a simple extension of the previous thermometer program that captures each reading in a list called TempRecords.

The current program puts a new temperature reading into Temp-Records as fast as mBlock can communicate with the mCore. The actual time this takes depends on the method used to tether the mCore to the computer. With the 2.4G serial adapter, the list adds about 340 entries in 120 seconds. With a wired USB connection, it adds over 500 records in the same time.

For a device that's monitoring the temperature in a hallway, even two readings a second might be overkill. Lists in mBlock don't have a hard size limit, other than the available RAM on the computer, so large lists aren't intrinsically bad. But to a human who might want to scan that list, three hundred entries of the same value don't add much to the picture. Using a small timer, we can easily add a 60-second delay between each temperature reading (shown in Figure 4-21). This ability also creates a powerful proving ground for investigations into size, scale, and sampling. These programs aren't answers to textbook problems! They're designed by humans to answer real questions in particular contexts. Given the physical reality of the problem site, is there a difference between two million readings taken every 30 seconds or one million taken every second? The correct answer for a drafty school hallway might lead to doom when applied in a high-altitude emergency shelter.

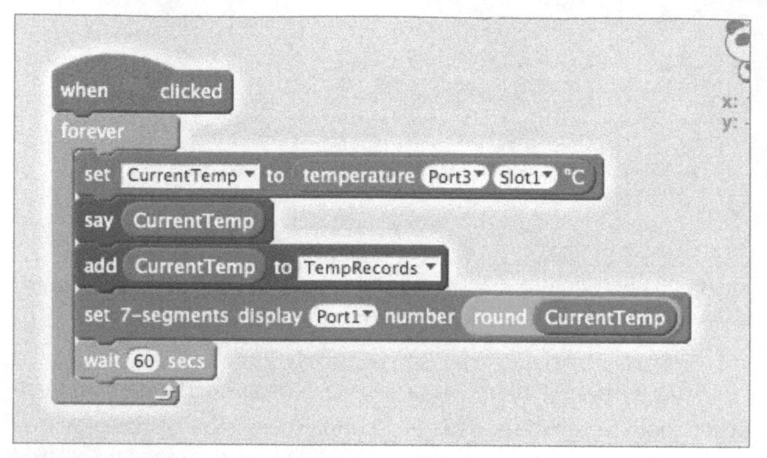

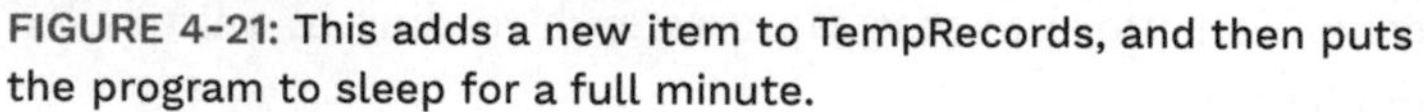
FIGURE 4-21: This adds a new item to TempRecords, and then puts the program to sleep for a full minute.

The choice to take a temperature reading every minute is clean, but a bit arbitrary. Why not 75 seconds? Why not three minutes? One of the powerful abilities you have in tethered mode is creating programs that can be modified on the fly.

Writing an adjustable program that uploads to the mCore introduces a number of problems: How do you know the current state of the program? What input method adjusts those states? What feedback is provided to the user to know they've successfully changed state? Trying to communicate the changing state of a program through a few blinking LEDs can be a nightmare. Tethered mode makes the screen, keyboard, and mouse accessible for any mBlock program, which allows you to create a clear, intuitive control scheme.

The first step is to replace 60 seconds, which serves as a threshold value for our loop, with a variable (see Figure 4-22). By itself, this doesn't change the behavior of our program.

Since SampleDelay is set to 60, that's the value check each time the program runs through the If statement. Now we can use mBlock's keyboard input functions to control the value of SampleDelay.

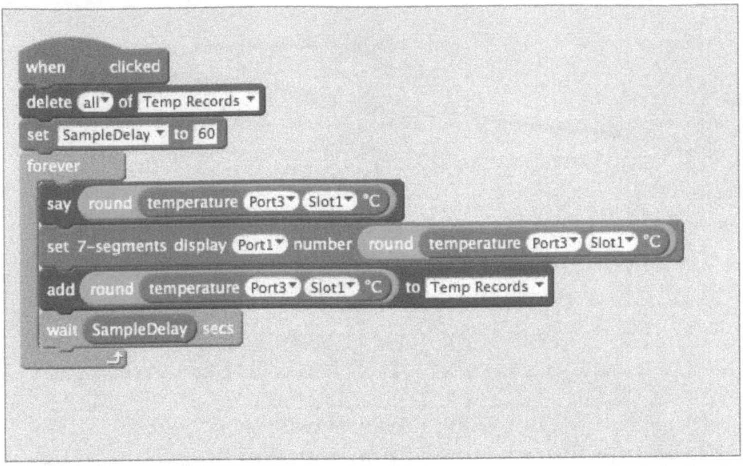

FIGURE 4-22: Setting SampleDelay before the loop ensures that the program will start with a 60-second delay each time it runs, and that value can be adjusted later.

These blocks make use of two different ways that Scratch looks for keyboard input. The first is the orange When Key Pressed hat from the Control palette (see Figure 4-23). When Key Pressed blocks constantly check for the given keyboard input, and then executes the blocks underneath once. However, if the signal stays on (like a keyboard key being held down) then the blocks will execute many, many times. To address this, the program waits for a pressed key to be released. This is a simple example of what's known as *debounce*—the process of eliminating stray inputs from physical systems. With this extra block in place, we're assured that each press of the up or down arrow will change the SampleDelay value by 5 seconds. Using the small checkbox next to the variable name or the ShowVariable block will ensure that the current value of SampleDelay stays on the mBlock Stage at all times.

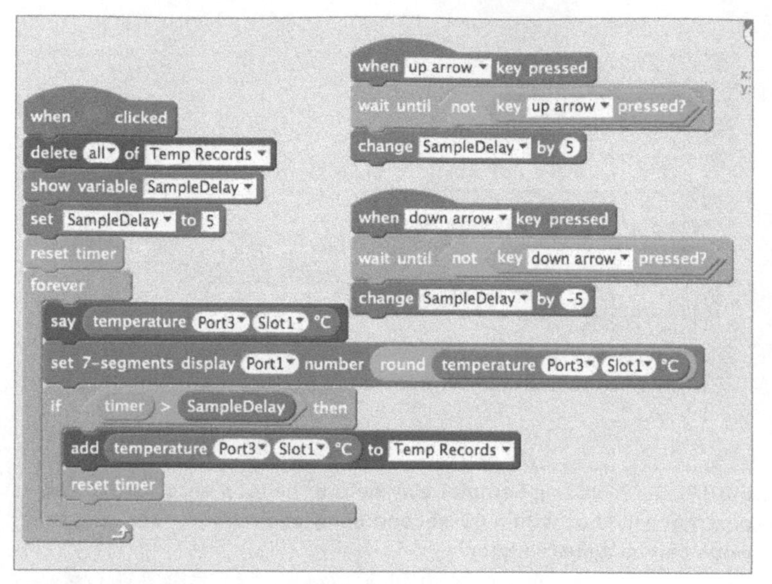

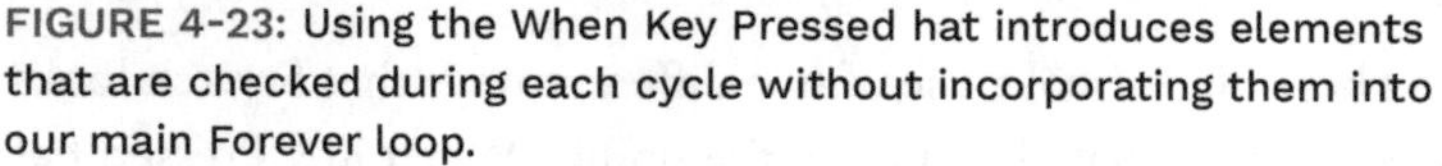

FIGURE 4-23: Using the When Key Pressed hat introduces elements that are checked during each cycle without incorporating them into our main Forever loop.

Once we've collected several hundred temperature readings, we need to find something to do with them! Creating data visualizations in Scratch-derived languages is a great, open challenge for new programmers. There are so many tools and hooks available in Scratch and mBlock that it's possible to create radically different graphs or visualizations from the same data. However, since those techniques are all grounded in Scratch programming, rather than the mBot, they're slightly outside the scope of this book. The book's website (*www.airrocketworks.com/instructions/make-mBots*) has a gallery of Scratch projects that create data displays from both static lists and incoming data streams. These projects can serve as models or inspiration, or even as a cautionary tale about what roads to avoid.

Fortunately, there's an easier way to get the data from mBlock into a spreadsheet using a tool that's purpose-built for handling giant lists of values. To export the values from any list in mBlock, you can right-click on the list viewer on the Stage and select Export (see Figure 4-24).

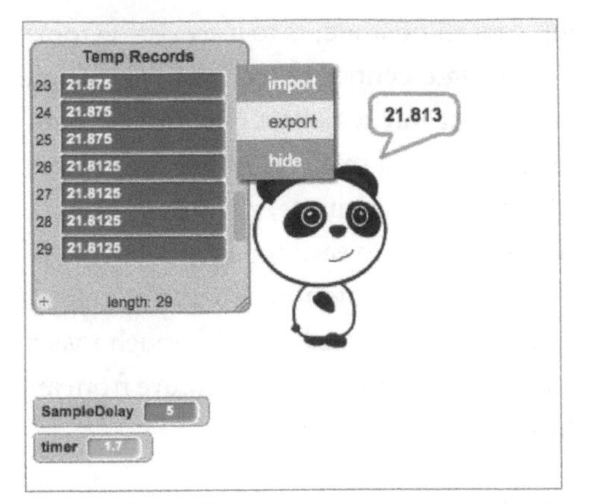

FIGURE 4-24: Export generates a list of return-separated values. Data in that format can also be imported into mBlock lists.

The list values are exported to a text file with each value on a separate line (see Figure 4-25). This format is perfect for copying and pasting the values into Google Sheets, Excel, Numbers, or any other comparable program.

mBlock temperatures

File Edit View Insert Format Data Tools Add-ons H

$ % .0 .00 123 Arial

	A	B	C	D
1	27.125			
2	27			
3	26.875			
4	26.875			
5	26.6875			
6	26.25			
7	25.875			
8	25.6875			
9	25.5625			
10	25.375			
11	25.25			
12	25.0625			
13	24.9375			
14	24.75			
15	24.625			
16	24.5			
17	24.375			
18	24.1875			

FIGURE 4-25: Data can also be imported into mBlock lists. List exports contain only raw data, so it's up to the user to add labels and context.

Not only does exporting data allow students to use existing tools to create simple graphs or calculate central tendencies like mean, median, and mode, but it's the best way to collect data from several mBlock programs. If several groups build small measurement devices to monitor specific areas of a larger problem site, exporting the data to a common document allows students to easily compare their findings. Consider three of these temperature monitors spaced out along a hallway. If all of their data is in a single spreadsheet, it's much easier to see how far and fast the momentary drop in temperature from an open door travels down the corridor.

DOOR MONITOR

While you can make interesting programming choices to measure and record temperature, the hardware didn't require much thought. As we've seen in this chapter, getting a basic reading boils down to plugging in the thermometer and finding the Read Thermometer block. The situation is quite different for more general tasks, like trying to track the open/closed status of a door. There's an abundance of sensors and methods that *can* perform this task, but none of them are called "door sensors," and mBlock doesn't come with a Door Sensor block. In this next section, we'll look at how to use a variety of tools to track door status, and how to design a program largely independent of the particular sensor.

When new roboticists approach a real-world problem like this, it's important to get them to look carefully at the physical details of the problem space. As a teacher, I've found that repeatedly asking basic questions about small observations and actions will eventually shift students' frame of reference down to a scale that robots and sensors can measure. It can take several minutes of climbing on a stepladder or lying on the floor while fiddling with the actual door to arrive at a sufficiently detailed answer to "*How does the door open?*" What makes these questions wonderful for groups of kids is that deep, detailed observation reveals that not all doors are the same! Differences in the material, frames, weight, and construction of each door will push students to generate novel solutions that accommodate all that messy, real-world variation.

For a sliding door, the mBot ultrasonic distance sensor was able to face inward and watch for a gap when the door opened. (See Figure 4-26.)

The mBot Line Follower sensor is placed near the hinged edge of the door. Popsicle sticks are used to enlarge the visible target of the door. The purple sticks shown in Figure 4-27 block one of the two light sensors on the Line Follower until the door is opened. This creates a unique closed state where the two different sensors on the Line Follower report different values.

FIGURE 4-26: Early versions of this sensor failed when kindergarten students would only slide the door open far enough to squeak through.

FIGURE 4-27: This popsicle stick solution emerged after students struggled to find a sampling rate that would reliably catch the door in motion.

These students mimicked common commercial door alarms and used a magnetic reed switch attached to the door frame and a magnet stuck to the metal door (see Figure 4-28).

FIGURE 4-28: One group of students latched onto reed switches early and searched the campus for a metal door to hold the magnet in place.

Reed switches have a small lever encased in the plastic housing that moves in response to strong magnetic fields. Reed switches work like any other button would with the mCore. On the RJ25 connector, connect one wire to the ground pin and the other to the S1 or S2 pin (depending on the port you use). Here, the bare wires from the reed switch are soldered to standard 0.1 pitch header pins, which match nicely with the JST connectors on the RJ25 board (see Figure 4-29).

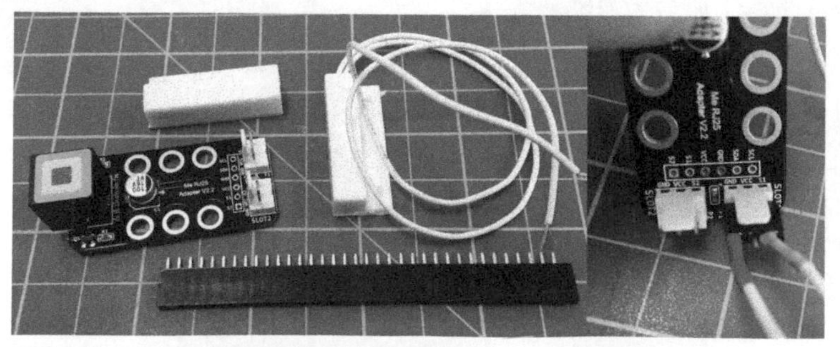

FIGURE 4-29: Soldering header pins to the RJ25 board make almost any crazy switch idea mBot-compatible.

While most reed switches are sold in packages with two plastic parts, the loose piece without wires just contains a magnet. In a classroom environment, this part often disappears and students wind up using generic magnets. Our Makerspace tends to use strong, rare-earth magnets due to size and storage concerns, but if you have access to a shoebox of red and blue horseshoe magnets, they'll work just fine.

This mess of tinfoil is a student-made button, where the foil-wrapped popsicle stick bridges the gap between two smaller foil pads, each connected to one pin of the RJ25 connector. (See Figure 4-30.) This is a crude button but gives decent readings using the same Limit Switch block as the magnetic reed switch.

FIGURE 4-30: Alligator clips are connected back into the RJ25 board with header pins, just like the reed switch.

NOTE It's worth mentioning again that these ingenious, messy kid-solutions are only viable because we use super-long RJ25 cables. If we were using the Makeblock-supplied lengths of cable, placing a switch at the top of an 8′ door frame would mean also mounting the entire mCore above kid height. When you use a 12′ or 15′ cable, masking tape can hold lightweight switches and sensor boards in precarious spots, while the mCore sits in relative safety. Long cables, neatly routed along doorjambs and floors, allow kid-made sensors to stay in active use for days and weeks without being a hazard to normal school-day navigation. Don't overlook the simple transformative power of extra length!

These solutions should not be seen as an exhaustive list of how to check a door. Simple problems often spawn complex solutions. It can be difficult for teachers and mentors to resist presenting the "right" solution, especially when kids are climbing a ladder to build something stupendously inefficient. Be strong and stay quiet! No matter how outlandish the solution, we can use the Custom Block features in mBlock to abstract away the mess.

Our goal with a custom block to check and open the door is that you could write a program that uses data from the door sensor without knowing anything about the physical construction. Instead of looking at readings from a particular sensor, we'll create a new variable in our program called DoorStatus and assign a status of either Open or Closed. Variables in mBlock can store letters or numbers, and can perform type-appropriate operations on them. Subtraction doesn't work well on text (or more properly, *strings*), but mBlock can check for equality. For strings, *equality* means an exact character-by-character match. There's no functional difference between using Open/Closed, True/False, or 0/1 as DoorStatus values, but using Open/Closed makes the program far more legible for other humans.

We'll change the DoorStatus value only from within a custom block called CheckDoorState (see Figure 4-31). This way, the main program doesn't need to care whether the door sensor uses a Line Follower sensor or a magnetic reed switch.

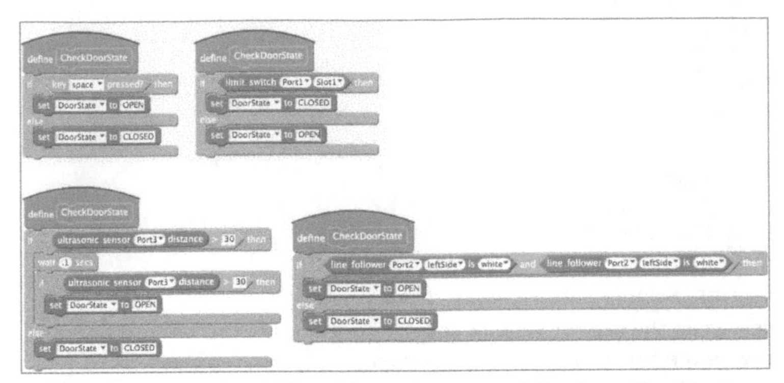

FIGURE 4-31: You should be able to drop any of the four CheckDoorState functions shown in these images into the same program and they should function identically.

All of these CheckDoorState blocks work for their particular sensor and physical setup, but the main loop doesn't need to know anything about those details. Using the CheckDoorState block means that the program trusts that, whatever happens inside, the code will update the DoorState variable accurately and promptly.

Using the same techniques from the temperature records program, this program tracks how long the door is held open in a list called OpenLength. This program uses two light blue blocks to monitor time in a program: Timer and Reset Timer from the Sensing palette. In a Scratch program, the timer starts running as soon as the program window opens, regardless of whether any program blocks are executing. This value will increase constantly until the Reset Timer block is triggered, and then it will reset to 0 and start counting again (see Figure 4-32).

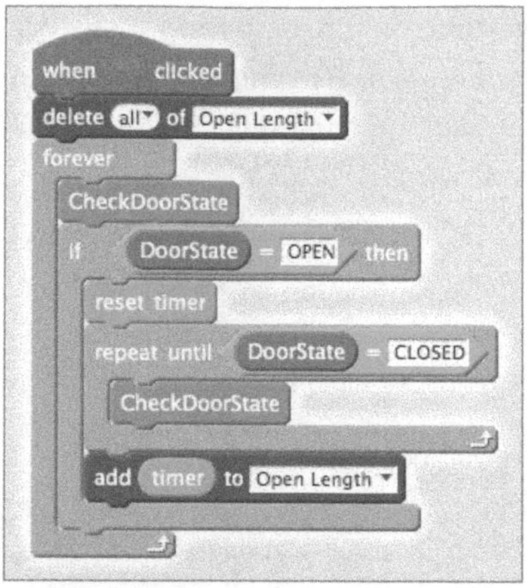
when clicked
delete all of Open Length
forever
CheckDoorState
if DoorState = OPEN then
reset timer
repeat until DoorState = CLOSED
CheckDoorState
add timer to Open Length

FIGURE 4-32: In mBlock, Reset Timer serves as the starter's pistol and resets the timer, which is continuously running, to 0.

The resulting list records the OpenLengths with the Timer block's default precision of thousandths of a second. (See Figure 4-33.)

timer 13.5
Open Length
4 1.966
5 1.2
6 2.298
7 5.8
8 5.266
9 2.067
10 2.267
11 1.332
+ length: 11
DoorState CLOSED

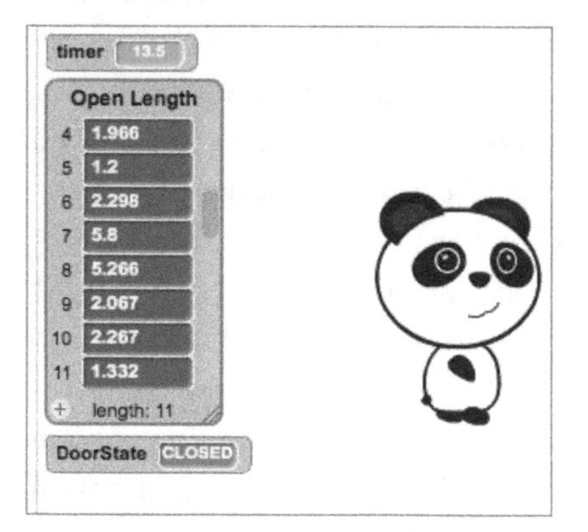

FIGURE 4-33: You can see that even though the door was closed after a second, the timer value is still increasing.

Unlike the record of temperatures, which took a reading at fixed time intervals, an entry is added to the list only when the door is opened and closed. As a result, the number of entries in Open-Length provides a count of how many times the door was opened. In mBlock, you can access that number directly using the Length Of List block from the Variables & Blocks palette.

Knowing that this value will go up every time a new item gets added to the list opens up the possibility of calculating averages or looking for patterns. it's a good practice to isolate extra tasks in custom blocks, as we've done with the doorstate. Strive to keep the main loop readable and the custom blocks narrowly focused. Using more custom blocks in a program, provided they're well named, reduces complexity for both the designer and the user (see Figure 4-34).

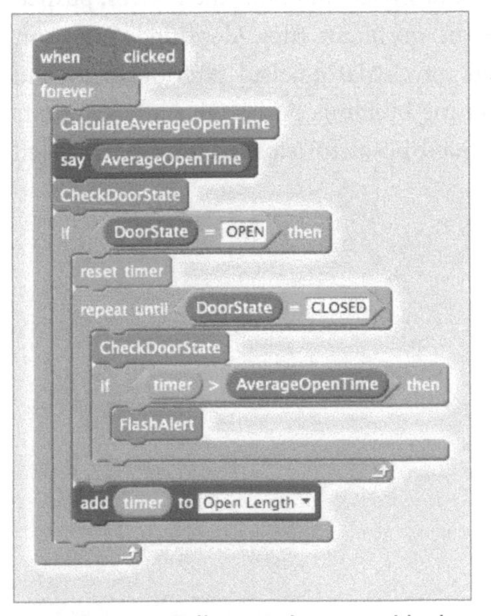

FIGURE 4-34: Well-named custom blocks create mBlock programs that are clear and human-readable.

Without seeing the scripts contained under these custom blocks, a reader can still intuit what should happen. Names alone suggest that the custom block CalculateAverageOpenTime will update and set the variable AverageOpenTime. When the door is open for longer than that value, it's time for FlashAlert. This script could play a sound, illuminate a strip of LEDs, direct a robot to pull the door closed, or turn on a hose. In some way, the script represented by FlashAlert wants to motivate nearby humans to close the door.

Encourage young programmers who are satisfied with their first prototype to add custom blocks to programs in this aspirational manner. A new custom block has no scripts attached and will do nothing when added to a program. Empty blocks work as placeholders for new features and help the programmer consider *when* and *why* to perform an action independent of deciding *how*.

Using the mCore, students were able to gather real data about the quiet details of their everyday environment. Quantitative information, from machines they designed and built themselves, pushed them to find ways to fix the problems they identified. They put up reminder signs, cleared gravel that blocked exterior doors, and changed traffic patterns in the building. Although these solutions were small, they still provided powerful closure for a student-led learning experience.

Robot Navigation

Robots can navigate in a variety of ways. Autonomous robots navigate using programs that allow them to follow GPS coordinates, or sensors that allow them to navigate in response to their environment. Robots can also be operated by the user using remote control. We'll look at several types of robot navigation in this chapter. We'll also look at two add-on packs for the basic mBot kit, which are available for about $25. The mBot add-on Servo Pack and Interactive Light & Sound Pack have many additional brackets, studs, M4 screws and nuts, beams, additional sensors, and RJ25 cables, along with a wrench. These items are all handy for many of the projects in this chapter. The add-on packs are available through Amazon or on the Makeblock website (*www.makeblock.com*).

ROBOT NAVIGATION USING KEYBOARD COMMANDS

Connect your mBot to your laptop or tablet using Bluetooth or 2.4G serial wireless and write the code shown in Figure 5-1. With a Bluetooth connection, you can now control your mBot from across the room using the arrow keys.

Before we get started, though, I'll share a quick note about speed. If the robot is moving too fast, drop the bot's peak speed from 255 to 100. But remember: resistance in the gearing and the weight of the wheel combine so that lowering the values for speed too much may make you unable to move an assembled mBot. We've found that speeds less than about 70 aren't strong enough to move the mBot from a dead stop, but if you give it a small push, it will keep moving forward.

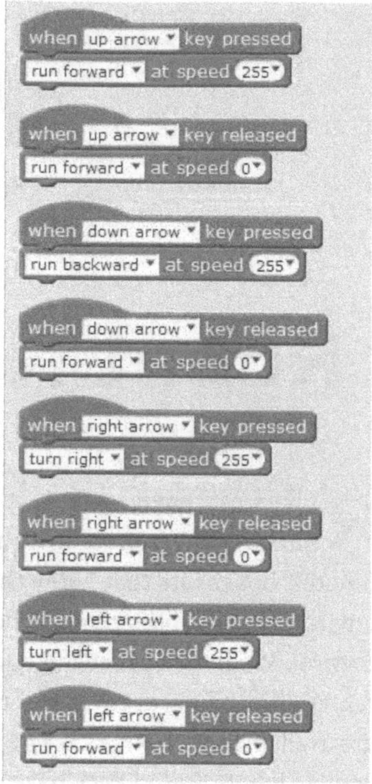

```
when up arrow key pressed
run forward at speed 255

when up arrow key released
run forward at speed 0

when down arrow key pressed
run backward at speed 255

when down arrow key released
run forward at speed 0

when right arrow key pressed
turn right at speed 255

when right arrow key released
run forward at speed 0

when left arrow key pressed
turn left at speed 255

when left arrow key released
run forward at speed 0
```

FIGURE 5.1: This Scratch code allows you to control your mBot with the up, down, right, and left keys on your keyboard.

Once you're able to navigate using your arrow keys, you can devise all kinds of challenges using other mBots. The signals to each robot won't interfere with each other, because Bluetooth connections are unique to each mBot. You can continue to use the code shown in Figure 5-1 in Scratch, but then add code to trigger actions based on input from other sensors. This way you can drive your mBot using the up, down, left, and right keys while your bot does other things.

ROBOTIC GAME CHALLENGES

Once you are able to navigate your mBot using the arrow keys, you can begin creating your own Battle Bots! A popular game is Sumo Bots, where two or more bots battle it out inside a ring marked on the floor using tape. The last bot inside the ring wins! My students came up with a great design using old CDs to "scoop" their opponent out of the ring, as shown in Figure 5-2. We'll start off with a few Sumo Bot defense and attack ideas.

FIGURE 5-2: Here is a creative take on a Sumo Bot challenge from some of my middle school students.

CD Scoop

You can add a CD scoop by attaching old CDs onto the front of the mBot just off the ground to scoop your opponent out of the arena.

Parts

Right-angle L brackets (2)	CDs (2)
4M×8 screw and nut (2)	Scrap wood

I'm trying to move my students beyond using gobs of tape for everything, so the following directions show the CDs being sturdily attached with drilled holes and screws.

1. Start off by attaching two right-angle L brackets to the front of your mBot using a 4M×8 screw and nut.

2. Glue two old CDs together with a hot glue gun, then flip the mBot over and, using the holes as a guide, mark the CDs where you'll drill a hole to attach them to the L brackets.

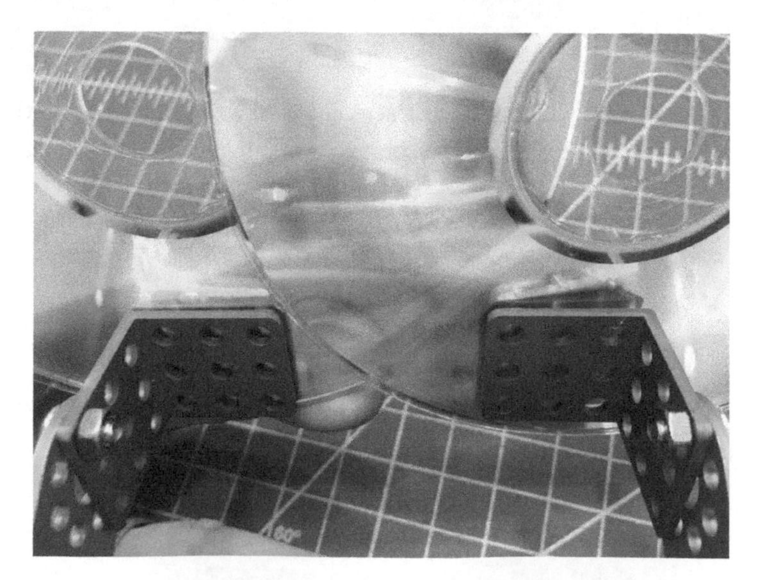

3. Place a piece of scrap wood under the CDs, and then drill through the CDs where you marked the holes.

4. Now put an M4×8 bolt and nut through the two CDs and tighten to hold them securely in place. Now you're ready to scoop your opponent out of the ring!

Spear-Lowering Servo

A BBQ skewer/lance that can be lowered using a servo can be added as an attack mechanism.

Parts

Bamboo skewer	4M×8 nuts and bolts
Mini zip ties (2)	9-hole plate
9g servo and servo holder	RJ25 Adapter
L brackets (2)	

1. Begin by attaching the two L brackets to the front of the mBot using four 4M×8 nuts and bolts, as shown in the following image. The L brackets, servo, and servo holder are included in the add-on Servo Pack.

2. Next, attach the 9g servo to the laser-cut acrylic bracket using the small bolts and nuts that come with the add-on Servo Pack. If you're using your own servo, download and laser-cut the file at *www.airrocketworks.com/instructions/make-mBots*. If you don't have

a laser cutter, you can print the full-scale PDF and cut by hand using cardboard or thin wood.

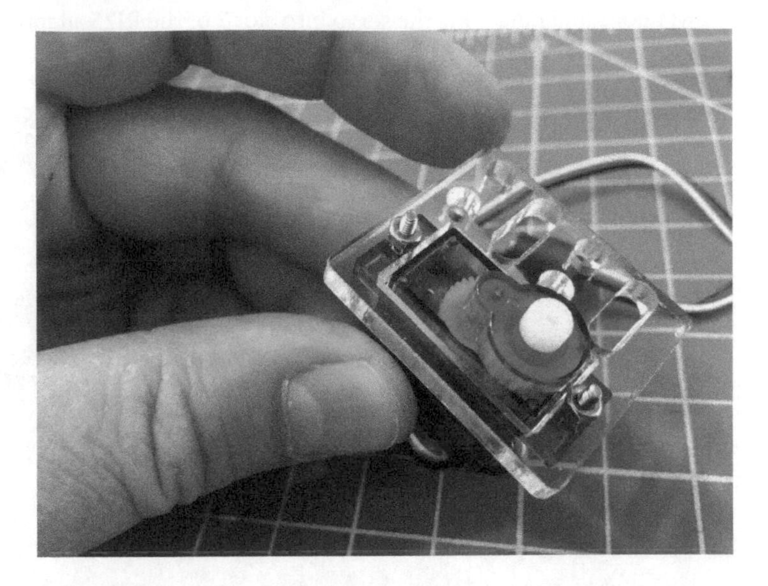

3. Remove the M4 screws from the back posts, which holding the mCore to your mBot, and replace with the M4×25 brass studs.

4. Now install the 9-hole blue plate to the brass studs with two M4 bolts, and then attach the RJ25 adapter to the blue plate with two M4 bolts and nuts. Plug the servo into slot 2 of the RJ25 adapter, and then plug the RJ25 adapter into port 4 on the mCore.

5. Before you attach the servo to the front of your mBot, you need to center it. Connect the mBot to your computer using your preferred method (Bluetooth or 2.4G wireless serial). Then write the code to center the servo, shown in the following image, and send it to your mBot. You can use this code with any program to center a servo.

Now you're ready to attach the servo arm to your servo using a very small Phillips head screwdriver and the tiny self-tapping screw that came with the servo.

6. Attach the servo to the right angles mounted on the front of the mBot using two M4×8 bolts and nuts.

7. Line up a bamboo barbecue skewer with the servo arm and attach with two mini zip ties.

8. Pull the zip ties very tight, and then clip off the ends of the zip ties with wire cutters.

9. In Scratch, write the code shown in the following image and send it to your mBot.

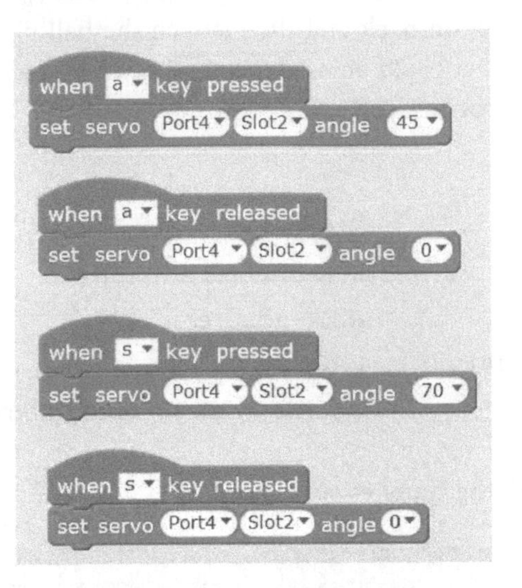

I had to edit the angles a little bit to get the lance to a 45° angle, and then a 90° angle. The spear will lower to a 45° angle when the A key on a keyboard is pressed. It will return to the up position when the key is released. The spear will lower to a 90° angle when the S key is pressed, and then return again to the up position when the key is released.

Now you're ready to joust with a cool skewer that can be raised and lowered using your computer. Safety is always important, so remember to wear your safety glasses when you're working with sharp things.

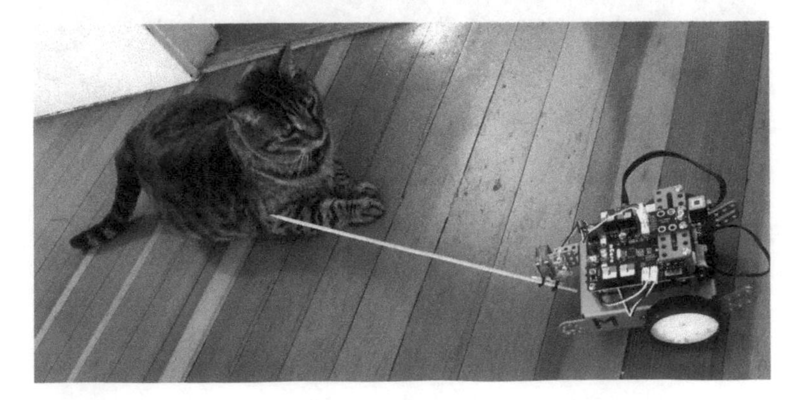

Catapult Ball Launcher

A whole different type of challenge is created using a plastic spoon and servo to hold the spoon back and then launch the ball like a catapult. This ball launcher could be used to knock down obstacles, shoot at targets, or aim for baskets.

Parts

Ping-pong ball	Staple remover
Plastic spoon	M4×14 bolts and nuts
Clear acrylic plate	RJ25 adapter
M4×25 Brass studs (4)	L bracket
9-hole blue plate	Double-wide 10-hole beam

Figure 5-3 shows all the supplies you'll need.

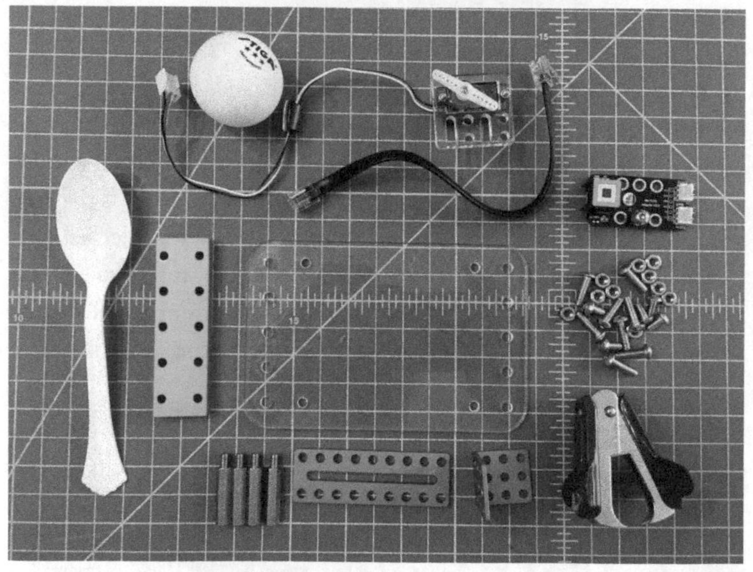

FIGURE 5-3: Here are all the supplies you'll need for a ping-pong ball launch rig.

The clear acrylic piece shown in Figure 5-3 was used in Chapter 1 as a case for the mCore. Here, we'll be mounting this platform to the top of your mCore to support the catapult mechanism and hold the electronics. Laser-cut files for the acrylic platform can be downloaded at *www.airrocketworks.com/instructions/make-mBots*, or printed out full scale from a PDF as a template for hand-cutting a material of your choice. Other key parts include a stiff plastic spoon and a standard staple remover.

1. Remove the four M4 bolts that are holding on the mCore, replace them with the four brass studs, and tighten securely.

2. Next, mount the acrylic platform onto the top of the brass studs using the four M4 bolts.

3. Install the 9-hole blue plate to the back of the plastic platform on the second and third holes from the left, as shown in the following image.

4. Mount the servo into the acrylic servo holder that came with the add-on Servo Pack following the instructions in the upcoming section, "Light-Emitting Head-Shaking Creature."

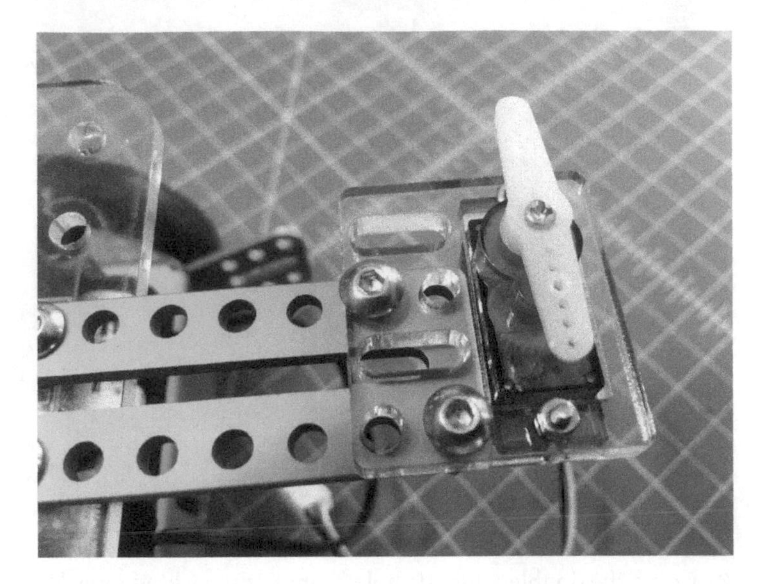

5. Next, use M4×14 bolts to bolt it down to the back of the 9-hole blue plate.

6. Screw the RJ25 adapter to the L bracket using two M4 bolts and nuts.

7. Using M4 bolts and nuts, attach the L bracket to the rear of the acrylic platform in the two holes on the far right.

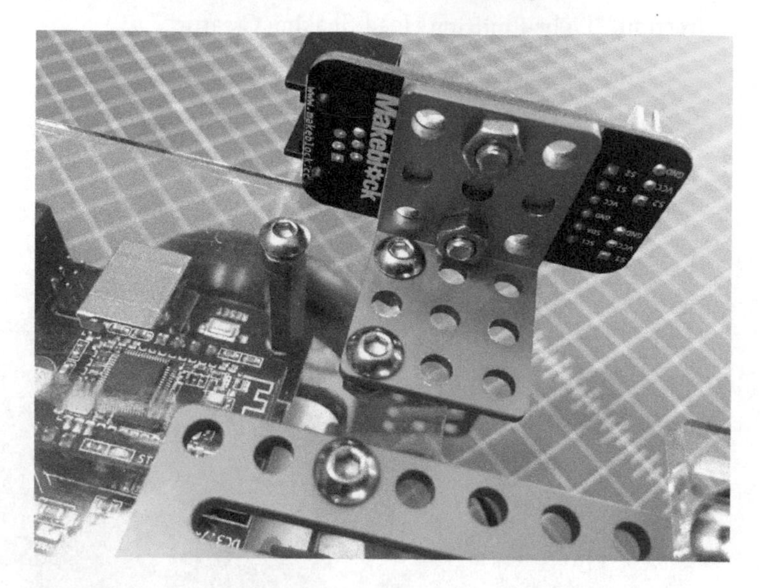

8. Cover the double-wide, 10-hole beam with masking tape. We're going to be hot-gluing the staple remover onto this part, so you'll want to protect the metal. Make sure you keep two parallel holes on the ends exposed, since this is where you'll attach it to the acrylic plate. It helps if you lay the masking tape on nice and smooth.

9. Now add a generous amount of hot glue to one side of the staple remover.

10. Press the staple remover evenly, glue-side down, on top of the tape, lined up with one end of the plate, as shown in the following image. Make sure the holes are exposed, leaving enough room for the M4 bolts.

11. Attach the staple remover assembly to the back of the acrylic plate with two M4×14 bolts and nuts, lined up as shown in the following image.

12. Test-fit the plastic spoon on top of the staple remover. The spoon should line up a little off center of the servo arm. The servo arm should be able to securely hold the spoon down in the trigger position. The servo arm will rotate out of the way, which will trigger the spoon catapult arm.

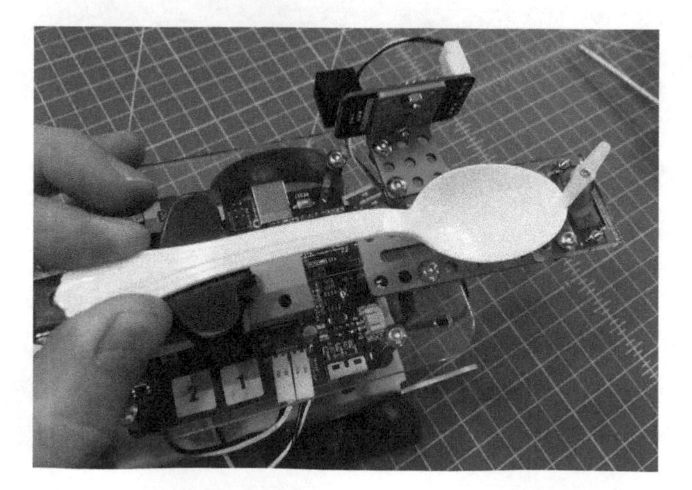

13. Once you know where the spoon should be placed (mark the end of the spoon with a Sharpie, if needed), add a generous amount of hot glue to the top of the staple remover and press and hold the spoon in place for 20 seconds.

14. Connect an RJ25 cable to the RJ25 adapter and to port 3 on your mCore.

15. Connect the servo to slot 2 on the RJ25 adapter. The nice thing about the servos that come with the add-on Servo Pack is that they only install in one direction so you always get them plugged in correctly. If you're using a generic servo, follow the directions in Chapter 3, "Head Turning Randomly Using 9g Servo and RJ25 Adapter." This is what your finished assembly should look like with the spoon catapult in the up position.

16. Cock the spoon back and rotate the servo arm in place to hold it. Place your ping-pong ball in the spoon and now you're ready to launch!

17. Create the code shown in the following image in Scratch. This code is really simple, with your space bar being the catapult trigger.

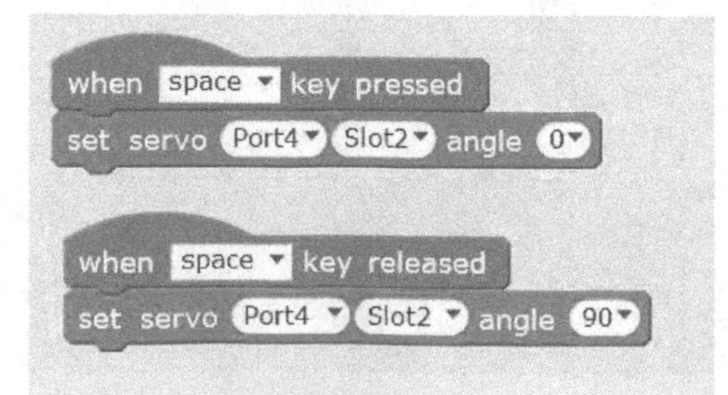

18. Next, test your code to make sure the trigger works. You may need to modify your code or center the servo (see step 5 in the "Spear-Lowering Servo" section for centering directions) to get your trigger to work properly.

Now go set up some targets or create some challenges and fire away!

Robot with a 9g Servo Grabber on the Front

For this project, we'll add an awesome 3D printed grabber mechanism powered by a 9g servo to the front of our mBot. By adding the grabber, which is controlled by your laptop, you'll be able to set up all kinds of challenges and even go head to head with other mBots to move items around a battle arena or obstacle course.

Printing and Assembling the Servo Grabber

Hats off to Jon Kepler for coming up with this brilliantly simple robotic claw and posting it on Thingiverse. Download it at *https://www.thingiverse.com/thing:18339* and print (printing will take about 35 minutes).

PARTS

- 3D-printed parts (as described in previous paragraph)
- 9g micro servo
- 3x8 mm machine bolt and nut
- Mini zip ties

Along with these 3D-printed parts, you'll need a micro servo (9g). The one shown in the following image uses metal gears but still costs only a couple of bucks. You'll need the servo linkage arms that go with the servos, a 3×8 mm machine bolt, and a

3 mm nut. Once you have all the parts printed and gathered you're ready to go!

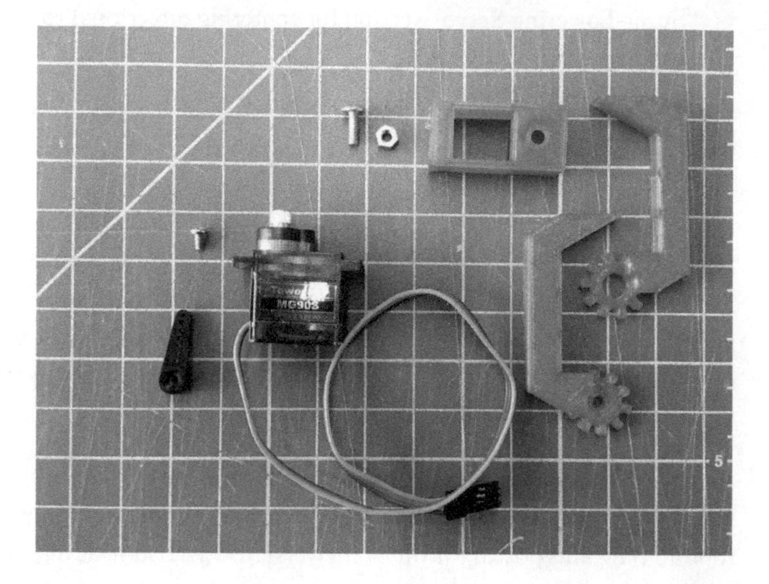

1. Turn the 3D-printed servo box over and push the 3 mm nut into the hex-shaped indentation.

2. With a pair of wire cutters, cut off the arm from the servo horn and then smooth out the cut edge with sandpaper.

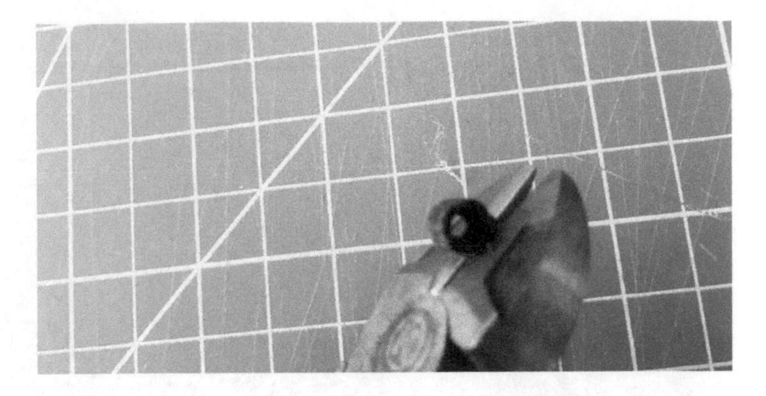

3. Place the servo box on top of the 9g servo, with the servo shaft positioned over the opening in the servo box.

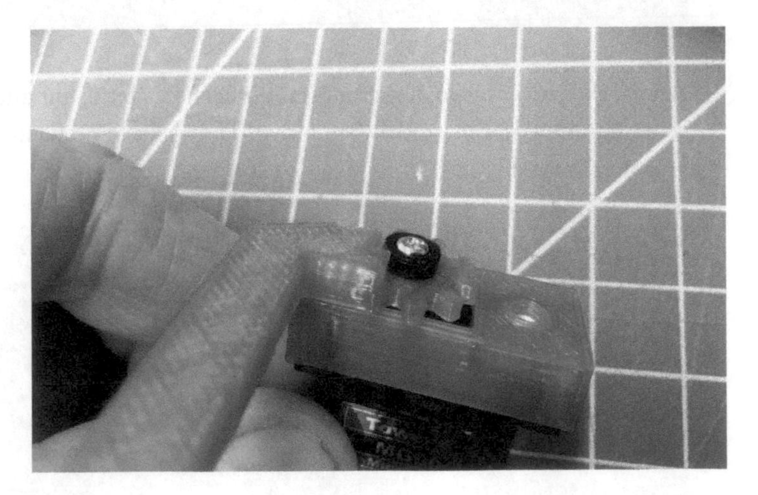

4. Attach the right pincer to the shaft of the servo with the screw that came with it. Use the piece of the servo horn from step 2 as a spacer.

5. Position the left pincer next to the right one with the gears interlaced.

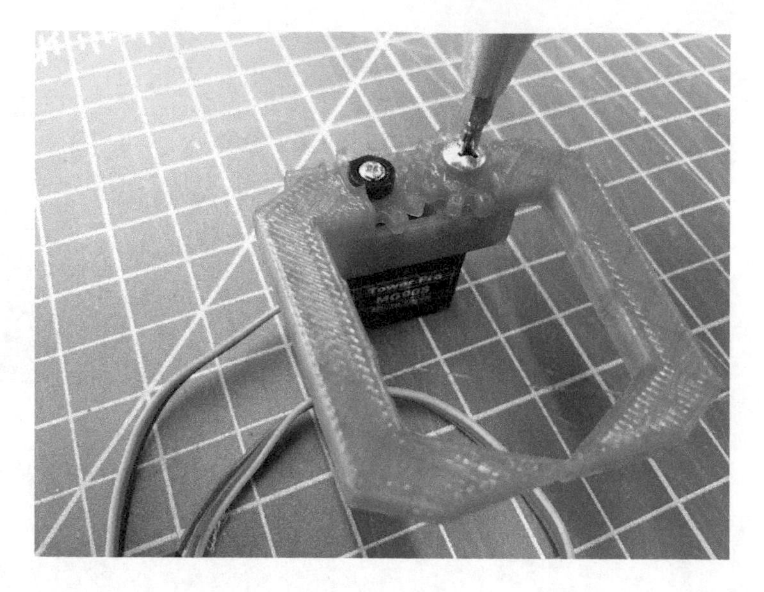

6. Attach the pincer by pushing the 3 mm bolt through the nut and tighten loosely so the pincers can move. They should move in a grasping motion.

The principle behind the servo arms is very simple. One arm is directly connected to the shaft of the servo. The other arm is linked by gears to the first arm. When the servo shaft turns, the first arm rotates and, thanks to the gears, forces the second arm to move in the opposite direction, thus bringing the two arms together. Once attached to the mBot, the arms may need to be adjusted after you get the servo calibrated.

Attaching the Servo Grabber to Your mBot

Next, you're going to build a bracket to attach your mBot to the grabber mechanism.

PARTS

L brackets	9-hole blue plate
M4 bolts and nuts (6)	Mini zip ties (2)

1. Screw the aluminum L brackets included with the mBot Servo Pack onto the front brackets of the mBot chassis using the M4 bolts and nuts.

2. Screw the 9-hole blue plate to the L brackets, as shown in the following image.

3. Attach the servo grabber to the front bracket with a mini zip tie, and cinch it tight.

4. Connect the servo to port 1 on the mCore using the RJ25 adapter. I attached the RJ25 adapter to the back of the mBot using some M4 screws and nuts. The wires can be neatened up using more mini zip ties or twist ties.

Here is the servo grabber in the closed position holding a piece of foam pipe insulation.

Write the code shown in the following image in Scratch. The code on the left controls the mBot using the up, down, left, and right arrows. The code on the right opens and closes the grabber claw using the space bar.

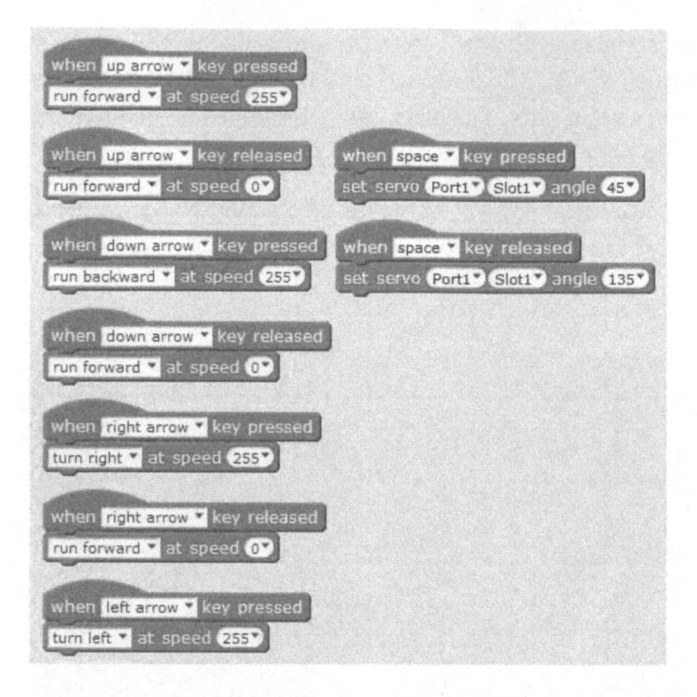

Light-Emitting Head-Shaking Creature

This project uses the add-on Servo Pack, which includes the following (also shown in Figure 5.4).

PARTS

M4 brass studs (4)	RGB LED sensor
M4×8 bolts	RJ25 adapter
M4 nuts	L bracket (2)
RJ25 cable (2)	Cuttable linkage (4)
Plastic spacers	M5 + M7 wrench
9g servo with holder	9-hole blue plate (2)

With the Servo Add-on Pack you can build a dancing cat, a head-shaking cat, or a light-emitting cat. For this project, we'll be combining the light-emitting and head-shaking features, which creates a robot with a lighted LED "head" that can move back and forth using a servo.

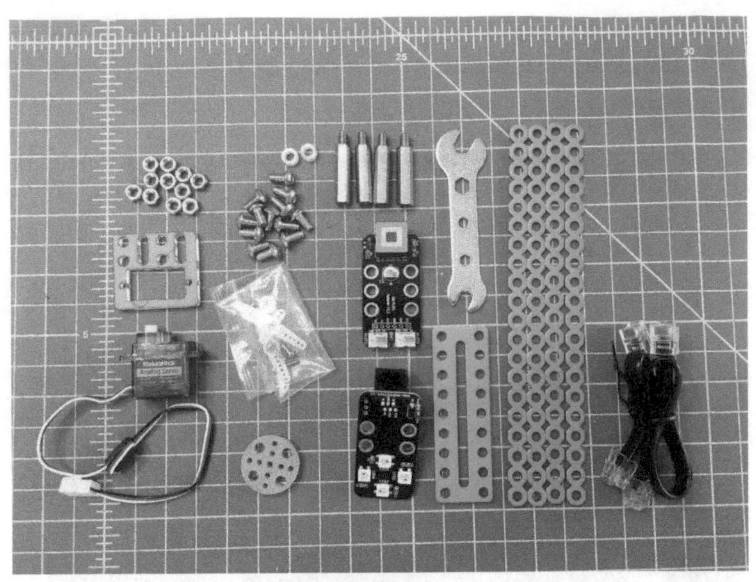

FIGURE 5.4: The Servo Add-on Pack

1. Attach the 9-hole blue plate to the top rear of your mCore.

2. Attach the RJ25 adapter and center the servo, as described in step 5 of the "Spear-Lowering Servo" section.

3. Once the servo is connected and centered, attach the L bracket to the servo arm using the two self-tapping screws that came with the servo.

4. Attach the LED sensor to the L bracket with two 4M×8 bolts and nuts, as shown in the following image. Make sure the sensor is attached in the top holes of the L bracket through the bottom holes of the LED sensor so that the sensor can rotate freely on the servo.

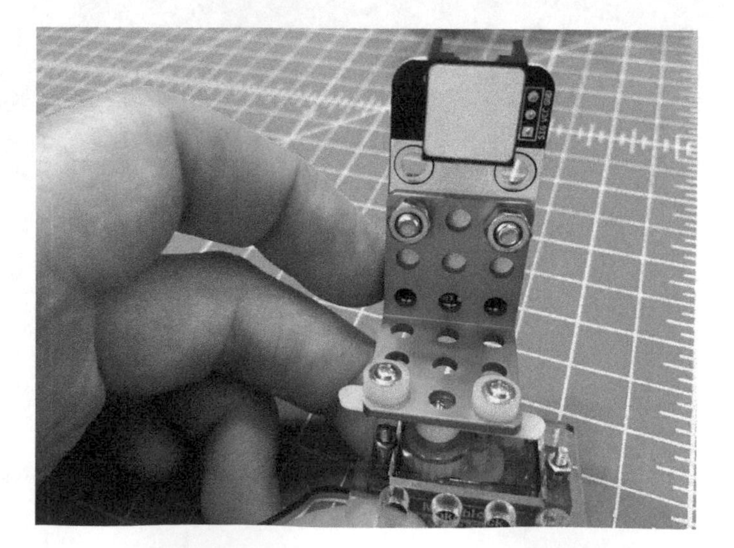

5. Plug the RJ25 cable into the sensor. It should be positioned so that it comes out of the top.

6. Plug the other end of the RJ25 cable into port 3 on the mCore.

7. Plug a second RJ25 cable into port 4 and plug the other end into the RJ25 adapter mounted on the back of the mCore.

8. Write the following code in Scratch. This will program the LED to turn on and off and move the light left and right using A and D keys, and re-center with S.

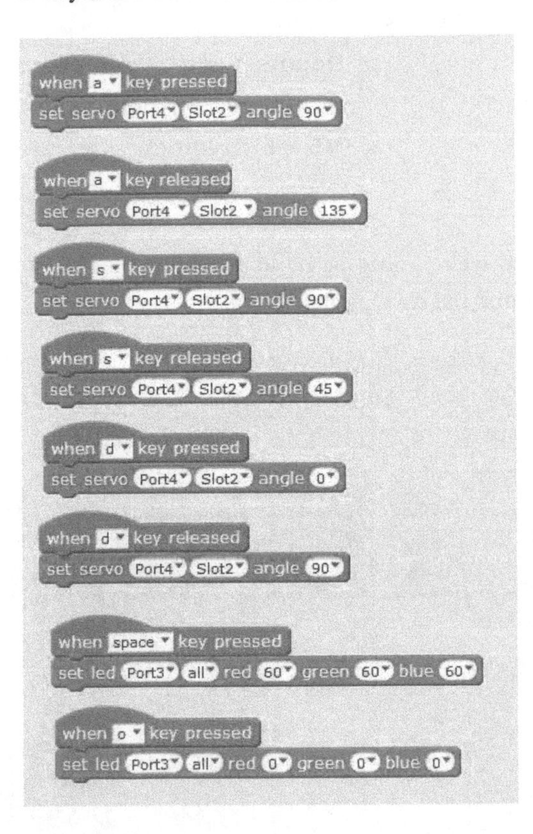

Light-Chasing Robot

For the following project, you'll use the add-on Interactive Light & Sound Pack. You'll be creating a bot that follows a flashlight using two Light sensors. The add-on Interactive Light & Sound Pack includes the following.

PARTS

M4 nuts and plastic spacers	45° metal plate
M4×8, M4×14, M4×22 bolts	Double-wide 10-hole beam (2)
Light sensor (2)	Double-wide 2-hole beam
RGB LED sensor (1)	Single-wide 5-hole beam (2)
Sound sensor (1)	M5 + M7 wrench
RJ25 cable (2)	

For this project, we're going to build the light-chasing robot using some beams and the two Light sensors.

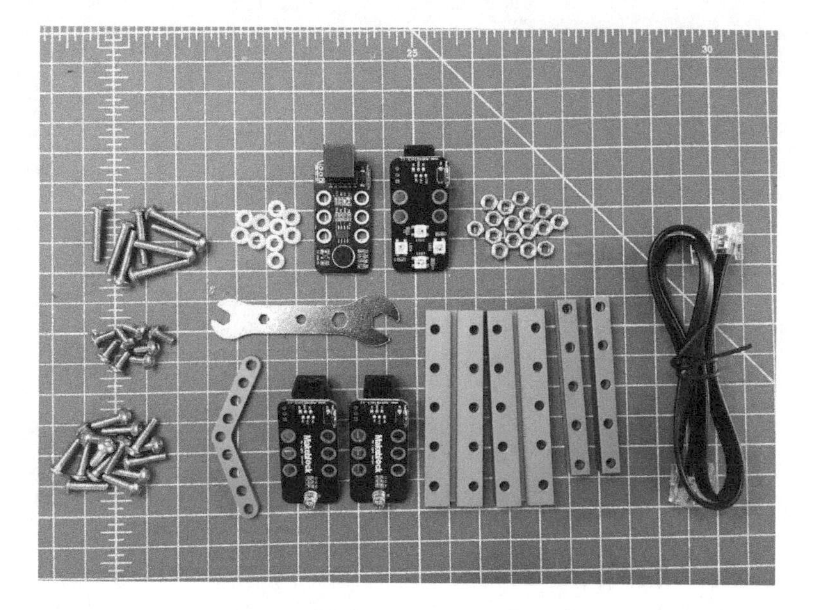

1. Mount the double-wide two-hole beam to each side of the front of the chassis with two M4×14 bolts and nuts.

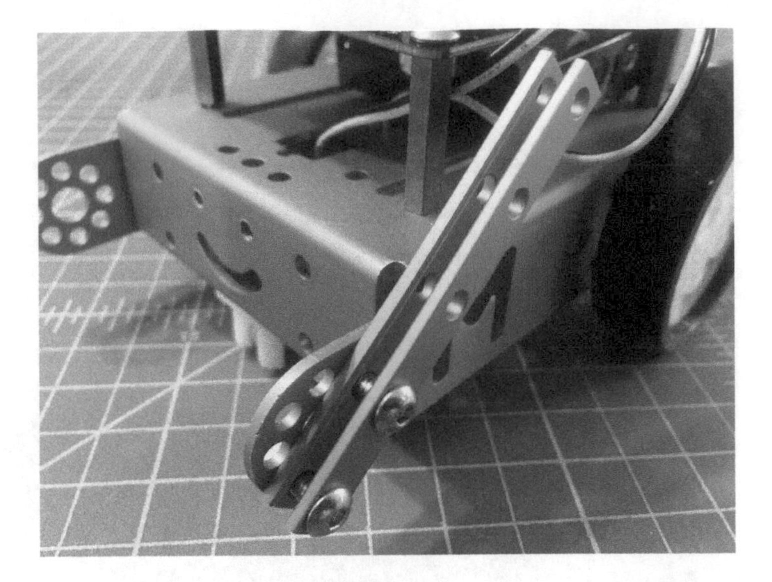

2. Each channel is threaded inside, so you can screw the Light sensor into the channel using two M4×8 bolts.

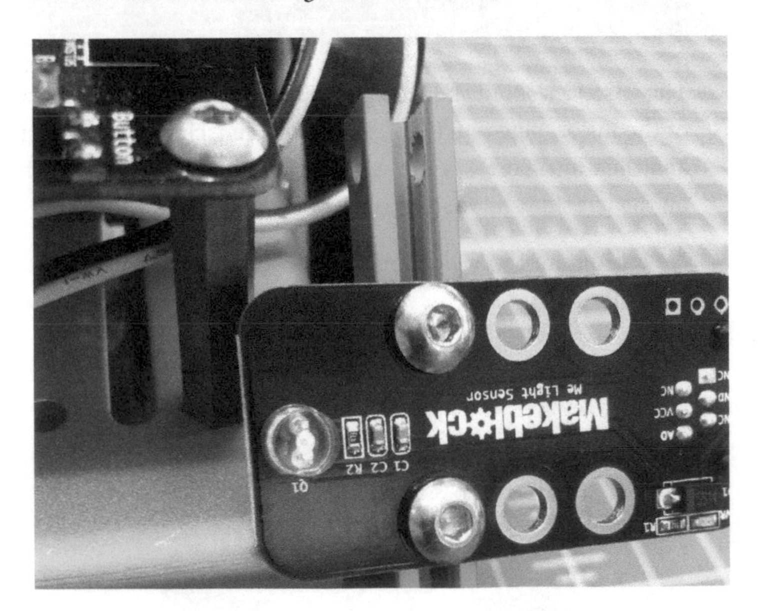

The following image shows both Light sensors mounted to the front.

3. The RJ25 jack should be facing out. As you're looking at the back of the mBot, plug one RJ25 cable into the jack on the right side and then into port 4, and then plug another cable into the jack on the left side and then into port 3.

4. Now program the following code into Scratch and send it to your mBot.

```
mBot Program
forever
  if light sensor Port3 > light sensor Port4 then
    set motor M1 speed 100
    set motor M2 speed 200
  if light sensor Port4 > light sensor Port3 then
    set motor M2 speed 200
    set motor M1 speed 100
```

You'll now have an mBot that follows the light from a flashlight, whether the light moves straight ahead, right, or left.

Maze-Solving mBot Using Standard Sensors

In Josh Elijah's Makezine.com article, "Beginner Robotics: Understanding How Simple Sensors Work," he describes the characteristics of true robots well: "For a robot to truly be considered a robot, it must be able to sense and affect its environment." The article uses a robot operation called Sense, Think, Act. In a nutshell, this means the sensor senses the environment, the microcontroller *thinks*, (makes a decision about what to do), and then it acts (carries out the decision).

The next project, brilliantly conceived by Dani Sanz from Spain (*juegosrobotica.es*), illustrates robotic operation excellently. His website is translatable using Google and I've translated his Scratch code here. Dani's project shows how globally the mBot platform reaches.

The Line Follower sensor and Distance sensor that come with the mBot kit are the only sensors needed for this maze-solving design. These sensors sense the environment, which in this case is a maze. The mCore thinks about what to do, and then carries out the decision. This feedback loop operates continuously from the time the mBot starts the maze until it finishes.

The mBot add-on Servo Pack comes with two L brackets, two plates, and plenty of M4 bolts and nuts, and they work well for this.

1. Using an L bracket, mount the Line Follower sensor vertically instead of horizontally (which is how it's used for line-following). Use one M4 screw and nut to hold the L bracket in place, and then add two M4 screws and nuts to secure the line sensor.

2. With two M4 screws, attach a 9-hole blue plate to the front right side of the mBot, pointing up vertically. Next, add an L bracket to the plate, facing out. Now, attach the Distance sensor upside down to the bottom of the L bracket facing out on the right-hand side of the mBot. Plug the Ultrasonic sensor into port 3 of the mCore.

3. Attach an L bracket to the front right of the mBot chassis using M4 screws and nuts. Plug the Line Follower sensor into port 2 and the Ultrasonic sensor into port 3 of the mCore.

4. Write the following code in Scratch.

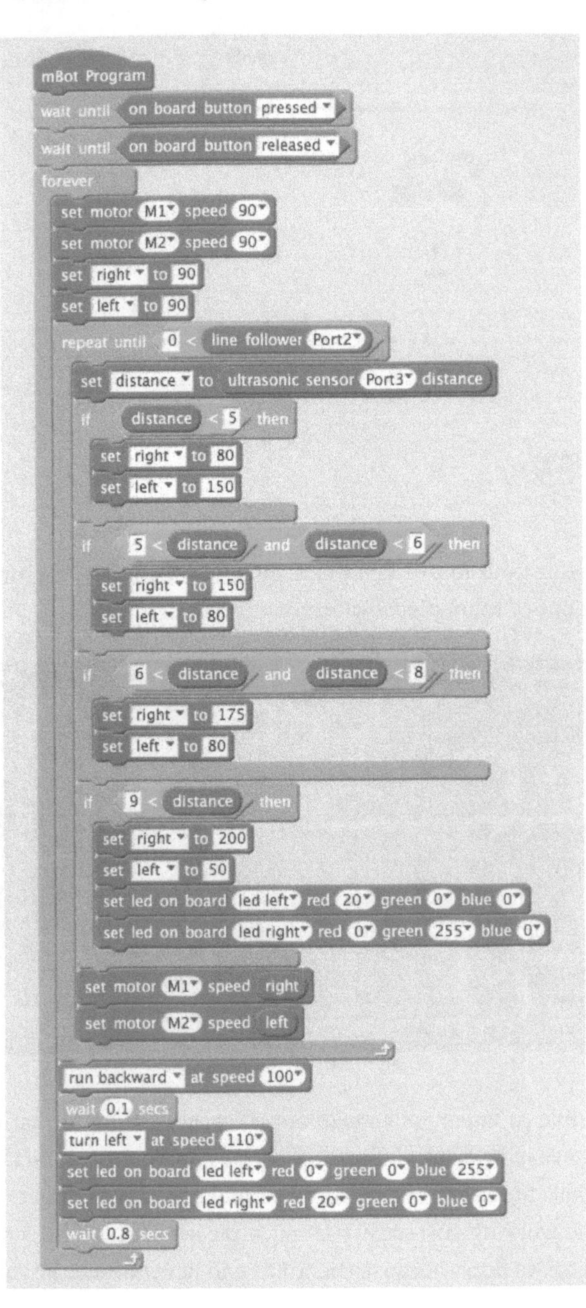

The following image shows the variables needed for the program.

5. Next, create your maze! The maze shown in the following image is made out of foam pieces placed on the floor.

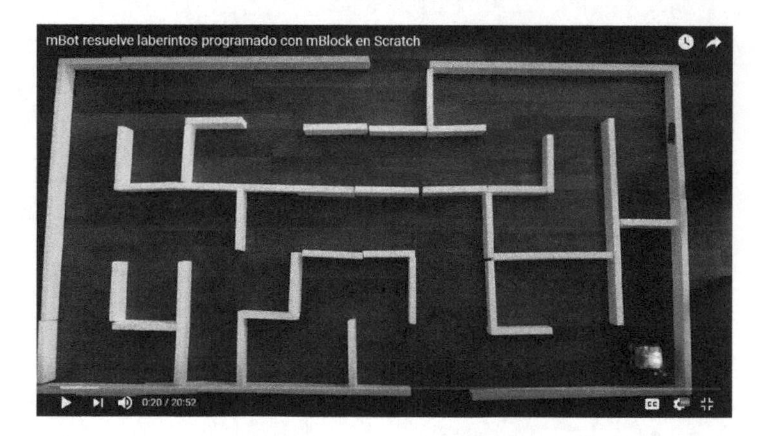

It's possible to make your maze out of cardboard, foam, or any object you have lying around. Start off with a simple maze, and then move the walls around and add more to make it more complex. If everything is working correctly, the maze pieces shouldn't need to be attached to the floor, because the mBot will never touch the maze walls. Kids will have a blast creating mazes for each other to solve using their mBots!

While this chapter has looked at many of the standard items offered by Makeblock like the add-on packs, the next chapter will really delve into how to use the mCore with off-the-shelf components like pumps, motors, and LEDs. Chapter 6, "Building Big and Small with mCore," will also dive deeper into the workings of DC motors and how to connect standard DC motors to the mCore board in a way that works with many projects and many kids.

Building Big and Small with mCore

This chapter explores the flexibility of the mBot through the frame of *dollhouse services*—designing simple and complex features, considered part of "smart" environments, on a smaller physical scale. However, the adaptability of the mCore platform means that it's entirely possible to scale up a clever idea from the dollhouse to the real world. On a small scale, we'll work with water, small LEDs, and servos, and then show how to adapt those programs to make use of household lamps, fans, and aquarium pumps.

We'll use the mCore to control several different devices. At the electrical level, these devices are mostly *two-pole motors*—sets of electrified copper coils pushing against magnets. The rotational force these components generate can be used to push water or air, or to spin a wheel or a propeller.

HARNESSING DC POWER

Brushed motors send currents to copper coils mounted on the spinning shaft, while *brushless motors* mount the coils on the stationary cylinder and spin a shaft covered with magnets. The differences in construction and scale can differ for particular applications, but the key point is that anything driven by a simple DC motor only needs a two-wire connection. Current flows through the motor circuit and generates spin. Most DC motors are non-polar, meaning that they will spin in either direction depending on the direction of electrical flow. However, if the DC motor is built into a fan or pump, the larger device may be built in a way that requires a particular polarity.

Servos, like the tiny 9g servos used in Chapter 5, "Robot Navigation," are geared DC motors combined with an encoder that reports motor position. The same is true for the LEGO EV3 and NXT motors. In each case, the encoders require extra wires to communicate their position back to the microcontroller. If you take a continuously rotating servo and only hook up the DC motor wires, you can use it as a plain DC motor.

Stepper motors consist of several sets of paired coils that each push the shaft a small fraction of a rotation (i.e., a *step*). These require more complicated control, normally in the form of a stepper motor driver IC, in order to fire each coil in a precise sequence and generate smooth rotation. Stepper motors are the backbone of 3D printers and laser cutters. Makeblock sells stepper motors and a Me Stepper Driver, but they're designed to work with Makeblock's larger boards, the Me Orion and Me Arguia.

DC motors are classified by a nominal voltage rating, normally printed somewhere on the motor body.

A 5V motor might spin at anywhere between 3V or 9V, but will work most efficiently at that 5V target. When the motor spins unencumbered, it draws a minimal amount of current. As the load on the motor increases, so does the amount of current it draws. This reaches a peak at the *stall point*, where the motor is under such load that it can no longer spin freely. Keeping a motor at the stall point for too long can burn it out along with the electronics in the motor control circuit. The mCore's design incorporates a small self-resetting fuse, which is essentially a tiny circuit breaker, to avoid damage to motors or the microcontroller. If any part of a circuit connected to the mCore draws more than roughly 1A, the fuse will overheat, trip, and cut power to the entire board. After a few minutes, the fuse will cool down and the mCore will power up normally. You should use those few minutes of inactivity to look into what caused the excessive load on the motors, and try to fix it for the next test.

Connecting Motors with Two Wires (Two-Pole Motors)

In theory, the mCore can control anything that uses a low-power DC motor, as long as you can connect the device to M1 or M2. But in real life, that last step is a doozy. Patching strange cables is a horrific time-suck, especially when you're working with a group of young people. In general, the "quicker" the solution, the more hours you'll spend later on fiddly repair.

One of the mCore's strengths compared with the basic Arduino is that it drastically reduces the amount of soldering and finicky breadboards. Even though breadboards are a time-tested prototyping tool, they don't stand up to kid use. In our Makerspace, projects are lifted in and out of project buckets daily, and occasionally get knocked to the floor. Soldering wires directly to the mCore would make more stable connections—provided you never wanted to use that board for anything else. No thank you!

The cheapest connector for the mCore's motor ports is a standard 0.1″ pitch header pin. We used this kind of connection on the RJ25 board when making simple switches. The long legs on stacking header pins are easier for anyone new to soldering.

1. To use header pins, trim a 2-pin section from the headers, and then strip both wires coming from the DC motor.

2. Put a smaller bit of heat-shrink tubing around the first wire, and then a larger diameter piece that slides further down and surrounds both wires.

3. Solder the first wire to one leg of the header, and then apply the smaller heat-shrink tubing.

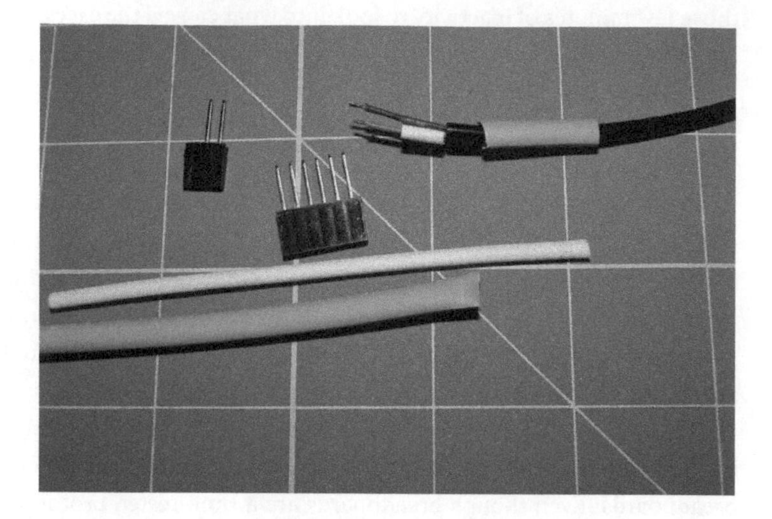

4. Now that you've protected the first leg, solder the other wire to the adjacent leg.

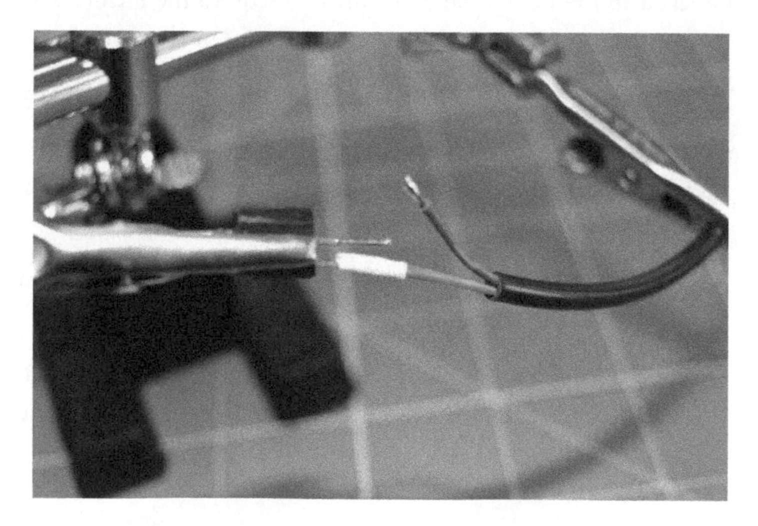

5. Apply the larger section of heat-shrink tubing, trying to capture some of the header pin's black plastic inside the heat-shrink tubing.

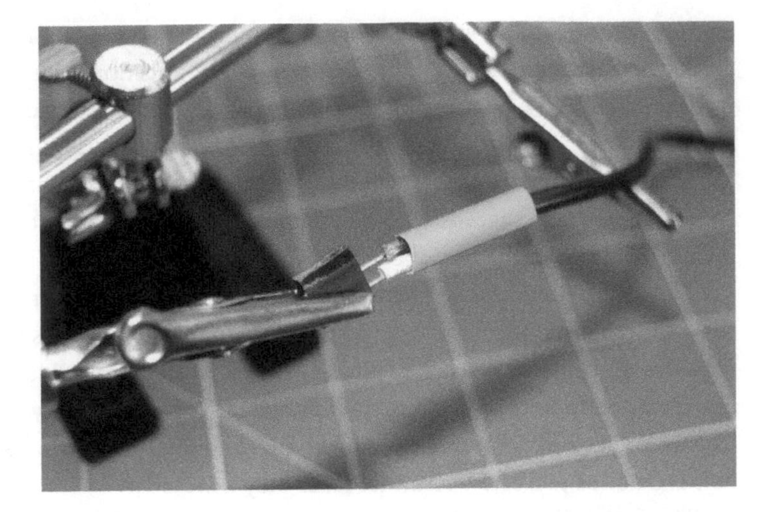

6. JST connectors have plastic rails to ensure that the plugs only fit on one way. These plugs, made of header pins, do not. Most DC motors will spin in either direction, so there's no damage if you accidentally plug something in the "wrong way." Mark the side of the header pins so that the positive and negative pins match the orientation on the mCore. The black plastic rejects most markers, but nail polish is visible and durable.

Soldering to header sleeves is good enough for a few motors, but represents a huge headache at scale. The connections might be solid to start with, but repeated stress can break them. If you need to make several connections at once, it's significantly faster to use a crimping tool and the appropriate JST connectors.

Adding JST ends or header pins to a DC motor works fine for connecting to the motor pins on the current mCore. If that's the only board you work with, then you don't need to worry about anything else. But Makeblock has shown remarkable inconsistency with connections across their current products. The Makeblock Ranger robot kit doesn't use the 2-pin JST connectors, and their external DC motor board uses a much larger 2-wire connector. In our Makerspace, two of the most common non-Makeblock motors in use are the LEGO NXT and EV3 motors. The sheer cost of LEGO components, and a low-level fear of future plug changes, drove my colleague Gary Donahue to find a more flexible connection system.

Gary's midpoint connectors have pairs of plug and socket connectors on one end with either breadboard pins or screw terminal connectors on the other. To make thse, we start by creating a large collection of small pigtails with JST plugs (which connect to the mCore) soldered to the breadboard pins of the plug.

Then Gary connects the socket end of the midpoint connector to the device cable. Since this is a screw terminal connection, it doesn't require soldering and doesn't permanently modify the motor. Making Gary's midpoint connectors requires a chunk of time, since you have to solder a large number pigtail connectors, but connecting a new DC device to a screw terminal takes only a moment.

The joy of Gary's midpoint connectors comes from the unexpected ability to reuse parts. When a student wants to reuse a DC motor salvaged from an old toy, Gary's system allows the kid to

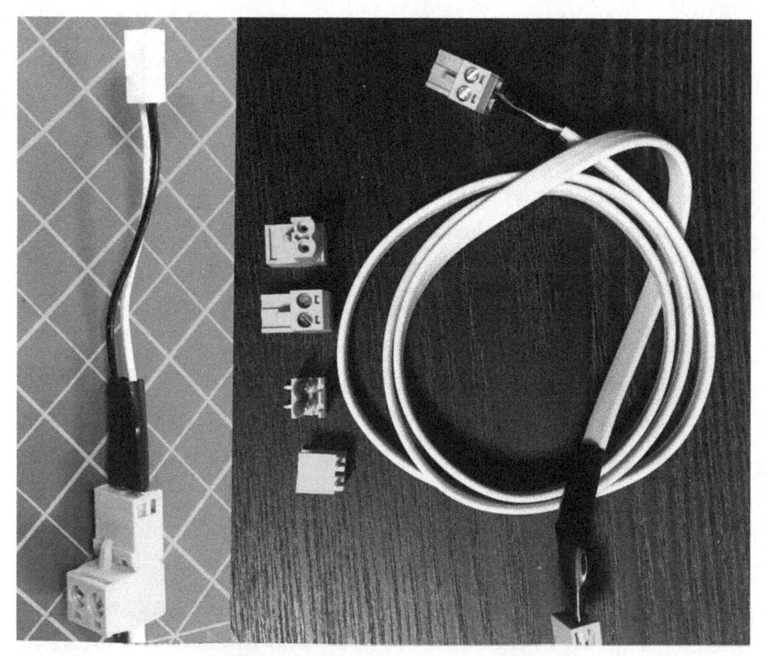

FIGURE 6-1: Short pigtails connecting mBot Motor pins to the common connector and long partner cables

"make the cable" by adding the screw terminal plug and connect it to their mBot in a few minutes. These connectors eliminate an incredible amount of cable-related hassle in our Makerspace, and allow kids to push their mBots in surprising new directions.

For the small-scale projects in this chapter, you can connect the extra fans, motors, and pumps any way you like. But if you're going to build two or three projects like this, take a lesson from Gary and invest the time in some midpoint connections.

BUILDING SMALL

Although we use the word *dollhouse* throughout this chapter, we avoid that term in classroom settings, because it may seem childish to certain audiences. In our Makerspaces, we use a variety of figures to match the scale of some projects. When kids are building complete environments, we'll look for 1–2″ figures, like LEGO mini figs or Playmobil figures. When designing clothing or furniture, 12″ poseable mannequins, bought for around $5 from IKEA, work wonderfully.

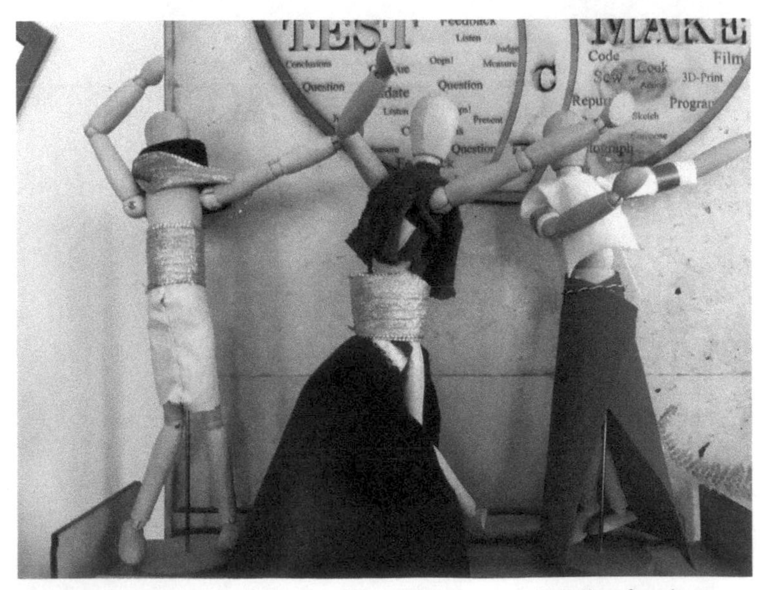

FIGURE 6-2: In class, we refer to anything we're building for these mannequins or another anchor figure as a *scale prototype*. Image courtesy of Chris Willauer.

Working on a fixed scale means that kids can move from idea to sketch to prototype quickly, without losing a moment in a long hunt for materials. Iteration is fast and cheap when you're making a parade float for 2″ figures or sewing a jacket for a 4″ torso.

For projects that create responsive environments, we've found that the best scale-human is the familiar LEGO mini fig. This way, an "apartment block" can be a vertical shoebox, single LEGO bricks can serve as furniture (see Figure 6-3), and clear tape becomes a useful building material.

Working in small scale lowers the cost (in time and materials) of "bad ideas," and ensures that students can have plenty of chances to learn from those productive mistakes. Miniature scale reduces the importance of detailed and accurate plans, something students struggle with and rarely see the value of. Instead of allowing a long time for planning, students can start to build their first prototype after only making a quick sketch. To help build planning and sketching skills, we ask them to make a careful drawing of their finished prototype, and then refine that drawing for the next build. Two small

FIGURE 6-3: A little LEGO work transforms a shoebox into a kitchen. Dishes in the sink are a nice touch.

steps, planning, then analysis, better mimics the Maker mindset, where the current work is always an approximation of the ideal.

Fire Management System—Small

With all of the aforementioned in mind, we're going to use the mBot to construct a fire suppression system for our dollhouse. We call it that because, without a narrative context, a segment of silicon tubing pumping water through a cardboard box doesn't *mean* anything. Twenty minutes of work, even shoddy work, can transform the same hardware into an apartment sprinkler system. We tap into every kid's imagination and diverse crafting skills by framing the project as a sprinkler system, instead of an abstract challenge of moving water between tubs. Within that framework, even simple decoration for the shoebox apartment requires choices that refine the scale.

The constraint of this prototype is that it must extinguish a fire that occurs on the stove. This constraint encourages builders to narrowly focus their work on the functional part of the system. As a prototype, this isn't better or worse than a "spray everywhere" sprinkler system, but adding that level of specificity allows students to leverage their real-world experience, so that each iteration of the cardboard prototype reflects and comments on that understanding.

Working with fire at any scale involves risk. In the small-scale shoebox apartment, even a single tealight could, if left unattended, result in a real and dangerous fire. In a classroom setting, you should limit the number of candles lit at any given moment. It's far easier to keep track of four flames than 40. Lighting and then quickly dousing the candle is the capstone of this project, but there's not much call for open flame before that moment. Along with the tealight, we'll explain how to use a wide-band IR LED to test the pump system.

DC water pumps come in several varieties, but for this project we've had the most success with submersible pumps. Makeblock sells the pump shown on the left in Figure 6-4, which has a nominal 12V rating, just barely within the mCore's power range. Unlike the submersible pump on the right, the DC motors and electrical connections on Makeblock's pumps need to be kept dry.

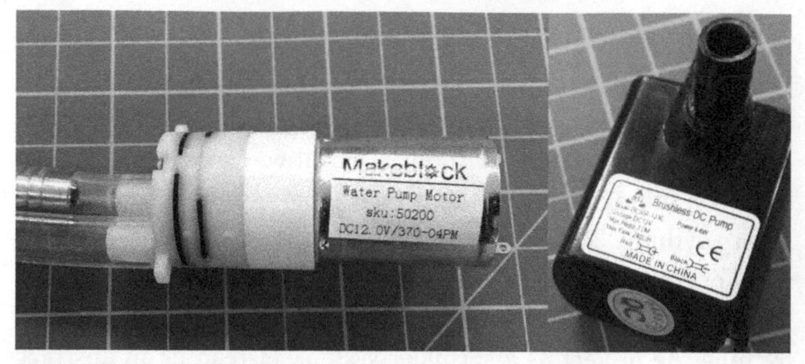

FIGURE 6-4: Makeblock's 12V pump just barely works on mBot motor ports and needs to be kept dry. The black submersible pump is a better choice.

In addition to the power concerns, we find dry pumps tricky to use in a group setting. Keeping an appropriate distance between the microcontroller, the electrical connections on the pump, and the flowing water requires a lot of space for each setup.

It's easier to find submersible pumps designed to work within the power range of the mCore's 5V motor supply. Searching online for "USB water pumps" will help filter out the larger aquarium pumps, which are too large for this prototype stage. Small submersible pumps are quieter than the external dry pumps, and only require a single outflow hose. Best of all, the pumps and the electrical connections are designed to be wet! We often have kids build a self-contained reservoir for the pump to be used throughout the prototyping stage, and that's what we'll built next. The parts for the reservoir are shown in Figure 6-5.

This example uses a glass jelly jar, but wide-mouth plastic containers would work just as well. First, punch or drill three holes in the lid. Holes for the water to flow through are sized so the plastic tubing fits in them snugly (see Figure 6-6). The hole for the electrical connector has to be large enough to accommodate the plug.

FIGURE 6-5: Here are the parts for the water reservoir. We'll make three holes in the lid: two for the tube and one for the power cable.

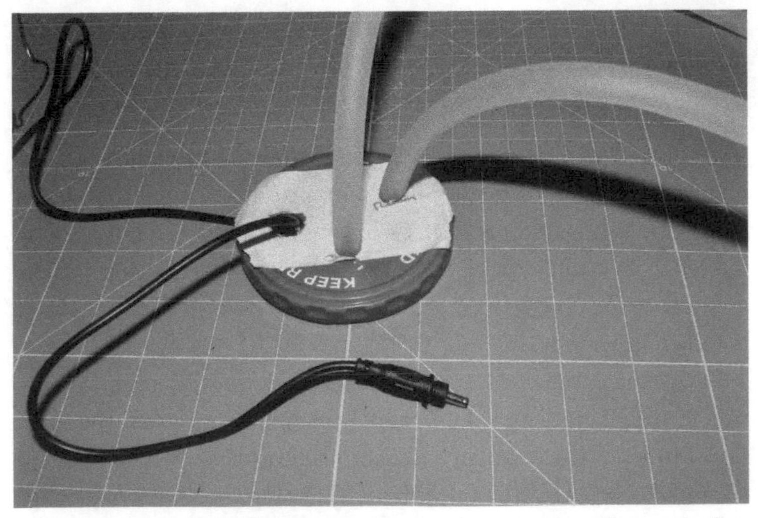

FIGURE 6-6: If the holes are too large, the tube may flop out of the lid when under pressure.

1. Connect one of the plastic tubes to the outflow nozzle on the pump. The return tube doesn't need to attach to anything (see Figure 6-7).

FIGURE 6-7: Connect one tube to the submersible pump's outflow, and let the return tube dangle.

2. Place the pump in the reservoir jar and pull out the slack on the power cable. If you're using a metal lid, be careful not to slice open the tubing (bad) or power cable (worse!) on a sharp edge.

Now you can fill the jar when the pump is in use and screw the lid and connectors in place. This contraption isn't spillproof, but it allows the pump and connectors to move around without soaking the work area. Blue tack or other moldable materials can seal the area where the tube travels through the lid. The completed reservoir is shown in Figure 6-8.

With the pump and water source secured, we will turn our attention to the flame sensor. Like most Makeblock products, the functional heart of the Me Flame Sensor is an off-the-shelf component mounted onto a small board with an RJ25 plug. The following image shows the Me Flame Sensor with the telltale RJ25 plug below and a similar component with a header pin connection.

FIGURE 6-8: Here is the completed water reservoir with the tubes attached and the pump on the bottom. You can use food coloring to help you tell from a distance when the water is flowing.

In general, what we call *flame sensors* are light sensors tuned to a particular wavelength of the infrared spectrum, normally between 760 and 1110 nanometers. Flame sensors actually combine an analog sensor, for numeric values, and a digital sensor that just reports whether there is fire or no fire. This digital reading also triggers a blue LED on the board and is controlled by a built-in threshold value, set by the small potentiometer (see Figure 6-9).

In our model, the sprinklers should only respond to an out-of-control kitchen fire. While an overzealous smoke alarm might be a kitchen annoyance, having a sprinkler set with too low a threshold makes a kitchen all but unusable. Managing the sensitivity of the flame sensor through physical placement and programming is the heart of this project.

There are some obvious concerns when kids are working with fire, but this project is a great way to mitigate those risks while enjoying the benefits. We use small tealights for the kitchen flames in these models, which provides more than enough fire to trigger the flame sensor. If placed too closely, it can also create enough heat to melt plastic tubing, crisp cardboard edges, or ignite stray paper scraps. Don't leave an open flame unattended!

FIGURE 6-9: Adjust the Me Flame Sensor's sensitivity by adjusting the potentiometer with a small screwdriver.

Anyone who moves a lit candle in and out of the model apartment risks wax-covered LEGO bricks and fingers. With regular attention, none of these problems threatens life or limb, and each one brings a useful "reality reminder" into the prototyping process. It's also possible to avoid these candle-related mishaps by using an IR LED "throwie" to test the position of the flame sensor.

LED throwies are a staple of classroom Makerspaces. Just place a 3V CR2032 battery between the legs of an LED, apply a little tape, and you've got a small light to stick just about anywhere (see Figure 6-10).

It may be the simplest circuit possible, but it delights and fascinates kids everywhere. But, since infrared light is outside the spectrum of human vision, it's harder to know that the light is really on. Make sure to place the longer leg of the LED on the smooth positive side of the battery and the shorter leg on the dimpled negative side. For anyone new to LEDs, it's helpful to do this with a visible-light LED at the same time (see Figure 6-11).

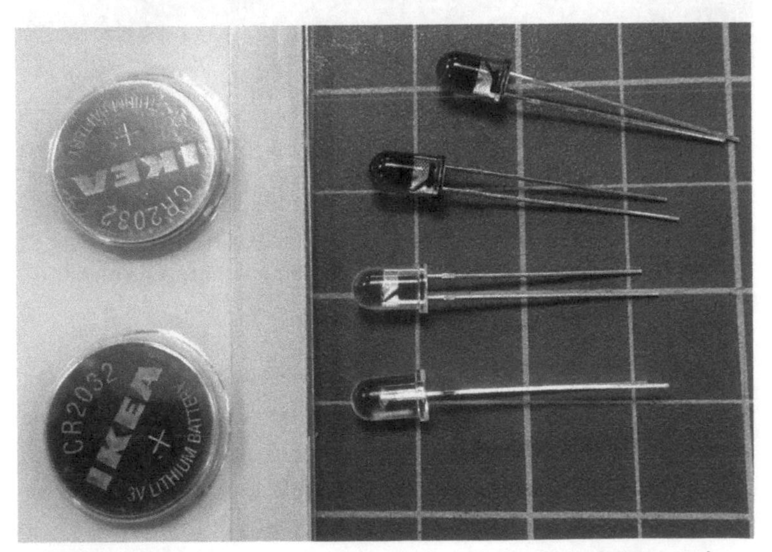

FIGURE 6-10: The top two LEDs use colored plastic to narrow and focus the IR light. Wideband LEDs with clear tops (bottom two) are better for this project but either would work.

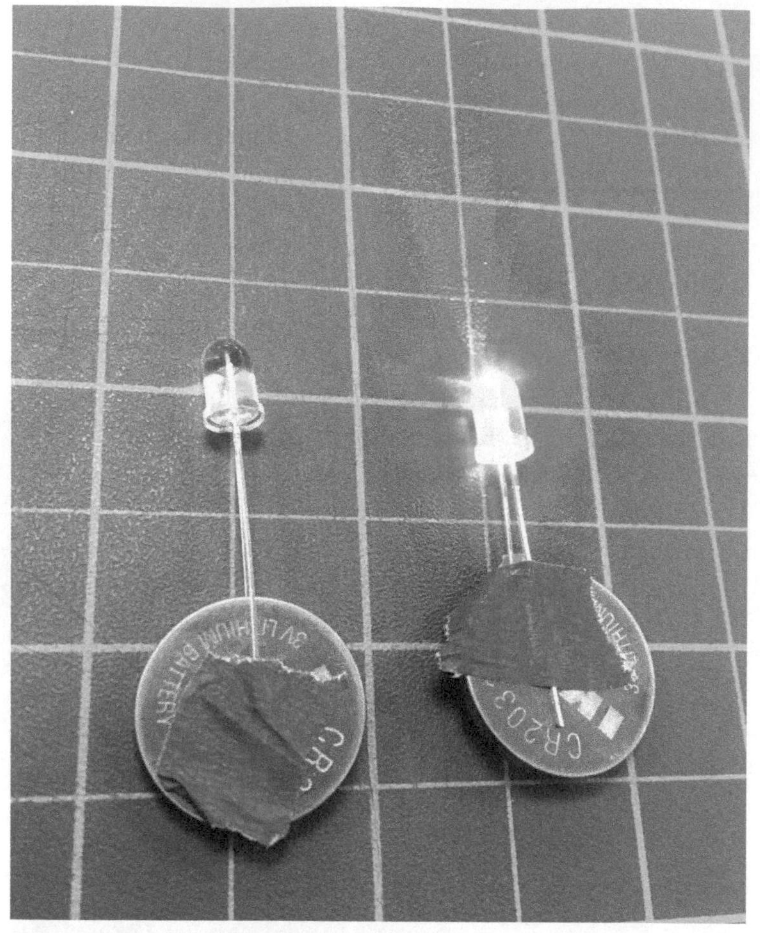

FIGURE 6-11: The wideband IR LED on the left is emitting as much light as the red light on the right.

Phone cameras used to provide a great way to check IR LEDs, since they capture a wider spectrum than the human eye. Today, the primary (rear-facing) camera on most phones uses software filters to clean up IR noise. Thankfully for us, that dubious feature hasn't yet migrated to the front-facing camera!

Since the flame sensor is actually an infrared sensor, these LEDs will easily impersonate a flame in a cardboard apartment. A large part of placing the flame sensor involves checking for ways the decoration and furniture might obstruct the sensor's view of the stove. While this could mean simply moving the sensor, many students will choose to adjust the candle or the stove instead. This leads directly to wax-covered fingers and other candle-related injuries. Using an IR candle instead of an open flame for these steps drastically decreases the risk of this project (see Figure 6-12).

FIGURE 6-12: This image shows bends a visible light red LED throwie into a more convincing tealight shape.

With our test candle ready to go, it's time to consider where to place the hose and sprinkler valve. This particular section of flexible tubing felt too large compared to the furniture, so we placed it on top of the cardboard box instead.

1. Mark the position of the tube on the outside of the box.

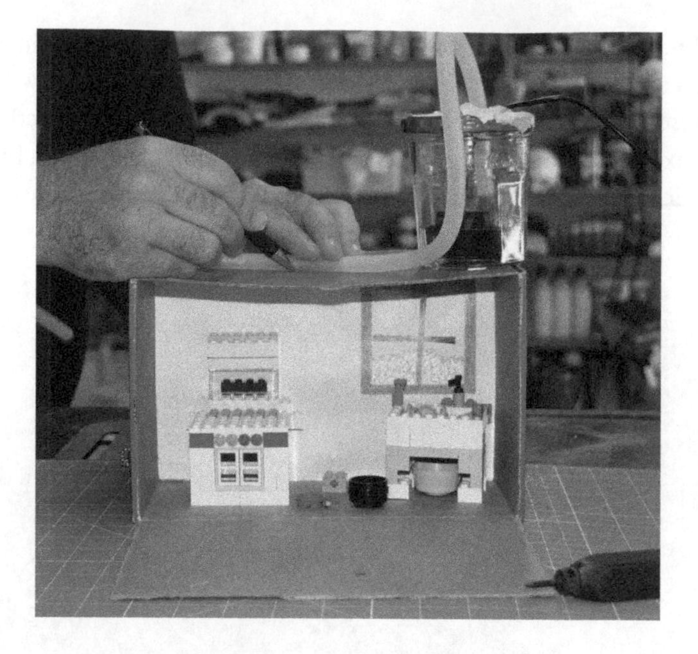

2. Place the tube so that it passes over the stove, then cut a small slit into the roof. It only needs to be large enough for water to drip through.
3. Use tape to attach the tube to the box on either end of the opening.

Even with the threshold knob dialed down, the flame sensor will spot a flame in the small prototype apartment from any spot with an unobstructed view. As an additional challenge, consider trying to hide the bulk of the sensor outside the box and make an opening for just the IR sensor.

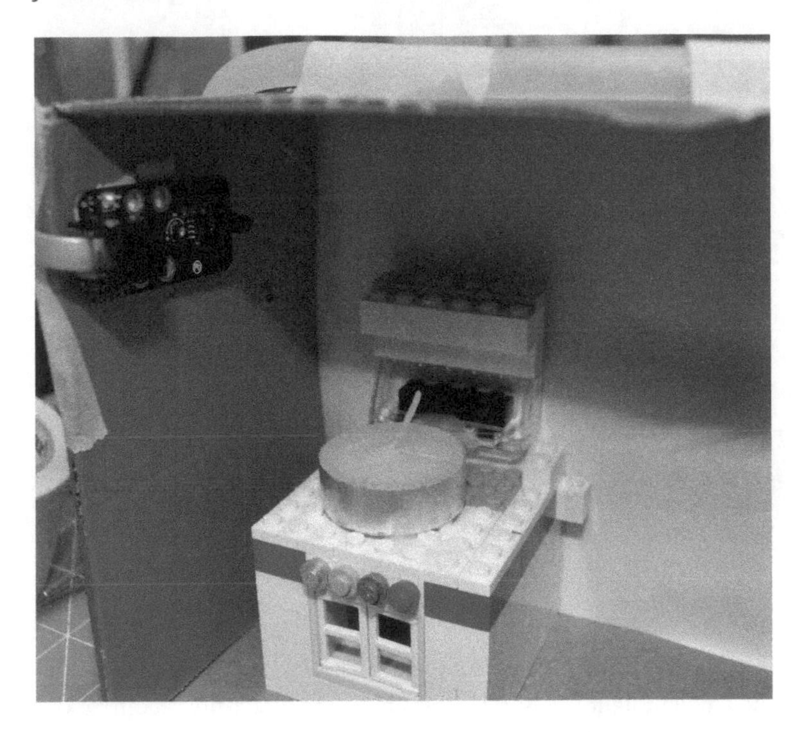

With the sensor placed, it's time to build the code, shown in Figure 6-13. The flame sensor has both an analog numeric output and a digital on/off output. The mBlock only reports the analog numeric value from the flame sensor. Use a Say block to check the sensor values as you move a lit candle or match in and out of the scene. When the sensor can see a flame, the number drops significantly. On the Makeblock flame sensor and most others, there's a small LED on the board that lights up when the sensor can see a flame. Keep an eye on this blue light while positioning the candle and adjusting the threshold knob to determine a useful threshold value.

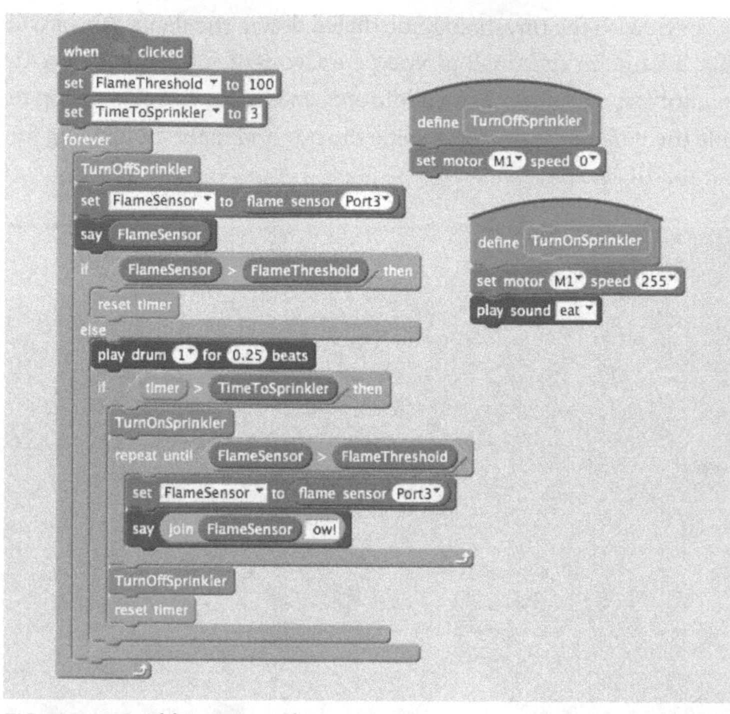

FIGURE 6-13: This program has many elements in common with the traffic light classroom volume meter from Chapter 2, "mBot Software and Sensors."

Our testing showed a flame sensor reading of 100 was well below the ambient light levels, but a bit above the direct fire readings. We used that for the FlameThreshold value at the top of the program so that it's easy to change, if necessary.

To avoid soaking the kitchen every time there's a momentary flare-up, this code includes a timing loop that checks how long there's been a flame in the kitchen. A single big flambé moment shouldn't trigger the sprinklers. The TimeToSprinkler variable is the number of seconds that we'll wait before turning on the sprinkler.

This program uses custom blocks to turn the sprinkler on and off. Since this action only requires one command block, it isn't saving any program length. Instead, it provides clarity in the main program, and flexibility if we change how the sprinkler connects to the mCore.

If the sensor reading is below our FlameThreshold value, the top part of the If/Else statement loops and resets the Scratch timer back to zero. However, once the sensor value clears the FlameThreshold value, the Else clause will execute and the timer will climb steadily. If the timer exceeds the TimeToSprinkler value, the pump turns on and will keep going until the FlameSensor value drops back below the FlameThreshold value.

We've used Sound and Say blocks to help track process throughout the program. When a program uses nested loops, it's tricky to say exactly what's being checked at a given instant. A drum sound plays every time the Else clause executes, providing an audio clue that the sensor is reporting a fire and that the timer is running. By adding a sound cue to the TurnSensorOn block, we can track any lag between the program's signal to start the pump and when we see water flowing in the prototype.

Now we're ready to test the system. Before any water starts splashing, we'll test the system as a closed loop (see Figure 6-14). Our pump will pull water from the reservoir, move it through the sprinkler system, and then back to the jar. This test is a useful practice

FIGURE 6-14: Here, we are testing the setup with a live flame and a sealed tube. Don't abandon the tealight in the kitchen!

for individuals, but crucial when working with groups. Ask that kids demonstrate a working closed loop system before they grab tools and poke holes in the tubing. Pierce the tube to add the actual sprinkler action when the code is solid.

This is the time to adjust the values for FlameThreshold and TimeToSprinkler. Our example uses about 1′ of tubing for the entire water path, so water reaches the kitchen less than a second after the pump turns on. For systems with longer hoses, such as one that has to reach a big bucket of water on the floor, it might be desirable to trigger the water a bit earlier.

Once those details are squared away, it's time to install the actual sprinkler! Use a felt-tip pen and put a small mark on the hose where it passes over the stove. Lift up the tube and drain the water from it. Use a hobby knife, scissors, or a pair of pliers to take a small notch out of the tubing (see Figure 6-15). When the tube is filled, pressure will force water out of this gap, so a small opening will work fine. A small snip is also easier to patch up with hot glue and electrical tape.

FIGURE 6-15: There's no need to remove lots of plastic. Water pressure will force water out of a small opening.

Replace the end of the hose in the water reservoir. The bulk of the hose should now be empty, so it won't drip into the apartment.

Here's the moment of truth! Start the mBlock program, then place the lit tealight on top of the stove (see Figure 6-16).

FIGURE 6-16: Success! Water from the sprinkler system completely douses the runaway fire on our stove.

Even though this cardboard kitchen won't last forever, we can make a few tweaks and test the system again before the box falls apart. Experiment with flame placement, with an eye out for spots outside of the splash zone that still trigger the sprinkler. Keep exploring ways to keep tiny apartment safe and intact.

Fan for Crowded Room—Small

The fire sensor demonstrates how a small-scale project with a reasonable narrative can support nuanced and complicated builds with very few components. It's also the best way to introduce and foster interest in projects that combine data from multiple sensors. Small environments are easily to monitor and manipulate, making it possible to mimic the automation in everyday life.

This project models common HVAC systems that engage fans or AC when the temperature exceeds a threshold, but only when the rooms are occupied. Don't forget to start with building the room—we can't overemphasize the value of asking kids to create and invest in the design and decor of the environment (see Figure 6-17) before they start working with the electronics.

FIGURE 6-17: A few fairies and some LEGO furniture can go a long way toward anchoring creativity and enthusiasm.

Our years of classroom observations suggest that decorating the test environment before starting a project inspires students, while making a "pretty box" for a functioning prototype feels like busy work.

You can use the same temperature sensor as you used for the sensor bots in Chapter 4, "Measurement Devices." This project serves as a natural follow-up to those data-gathering devices, and asks designers to use the power of observation to inform how and where to mount the thermometer. In a small-scale environment, sensors aren't even close to invisible when taped to a wall. Raising the imagined concerns of the scale-model inhabitants is a powerful technique for prompting and encouraging deeper thinking.

Since all small fans are just DC motors with plastic fins, use whatever materials you have on hand (three types are shown in Figure 6-18). Dollar stores often have handheld fans that use 2 AA batteries, which work well hooked up to mCore's motor ports. Makeblock sells an official version of this kind of fan, but it doesn't use the same connector as the mCore. Small electronics often contain 3V or 5V square fans, which are easy to attach to flat surfaces with tape. The littleBits fan Bit is one of this type, but unless you're already deeply invested in that product, it's not worth spending $15 on a $2 fan.

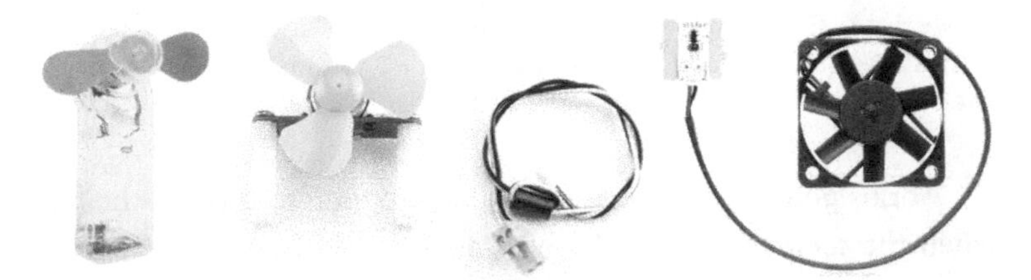

FIGURE 6-18: A handheld fan from the dollar store, Makeblock's official fan (its connector is shown to its right), and the fan Bit from littleBits

Unfortunately, the most common recycled small square fans come out of desktop computers and run off 12V. These fans normally won't move on the mCore's 5V power supply. If you have a large supply of these on hand, it might be worth using some of the higher voltage power techniques from the "Building Big" section later in this chapter to bring those into the mCore universe.

Whatever fan you choose, it's important to mount it in the scale room in a way that won't obstruct tiny feet or sever tiny heads. We've used a square fan and mounted it in the "window," as you can see in Figure 6-19.

FIGURE 6-19: A crowded apartment with a PC fan installed and a thermometer high on the adjacent wall

At this point, we can write a simple program to turn on the fan when the temperature climbs past a threshold (see Figure 6-20). Even though turning the fan on only takes a single block, it's worth defining those commands in custom blocks.

```
when clicked
set Temp Threshold to 25
forever
    say temperature Port3 Slot1 °C
    if temperature Port3 Slot1 °C > Temp Threshold then
        TurnON Fan
    else
        TurnOFF Fan

define TurnON Fan
set motor M1 speed 255

define TurnOFF Fan
set motor M1 speed 0
```

FIGURE 6-20: This is the State Check code from the sensor projects in Chapter 2.

It may require applying cold or hot fingers to move the temperature quickly above and below the threshold. Don't let the giant hands reaching in and out of the room break the narrative illusion completely! Look for ways this simple feedback loop would delight or annoy the tiny people relaxing on the couch.

With these barebones environmental controls in place, it's time to consider how to determine if the room is occupied. Similar to the way we determined whether a door was standing open in Chapter 4, there's no single "correct" sensor that will help us answer this question. Determining how to use an arbitrary input to decide whether people are in a room is a great brainstorming activity, if not an entire project. Pull a random sensor out of a hat and see what you can improvise!

This example uses the passive infrared (PIR) motion sensor, not because it's best, but as a way to showcase the particular challenges associated with binary output. A PIR sensor uses pyroelectric materials that actually generate electricity when exposed to specific wavelengths of light. Unlike the flame sensor, the PIR doesn't report a value relating to infrared light levels, but responds when that light value changes significantly.

In mBlock, the angled frame of the PIR block indicates that the sensor's values will always be either 0 or 1. When the PIR motion sensor block reports 0, that indicates that infrared levels have been basically stable for the last few milliseconds, which we interpret to mean that nothing large or warm is moving nearby. If the infrared level changes in that small window of time, the sensor reports a 1. The basic shorthand of 0 = still, 1 = movement, works for most situations, but it's worth considering the edge cases. A flickering candle will confuse a PIR sensor and it will report motion, whereas snakes and other cold-blooded reptiles could sneak by undetected—shudder. That is yet another reason to be afraid of snakes.

Most people already have an intimate familiarity with PIR-related frustration from countless restrooms. From finding the right position of your hand for automatic sinks, to stingy paper towel dispensers, many public bathrooms are crowded with PIR sensors. One strength of this project is that it can put kids in control of the same kind of robotic systems they encounter every day. As a general principle, humans should be able to build something as sophisticated as their own bathroom fixtures.

Since the PIR is a binary sensor, our program should look for readings over time rather than a single reported sensor value. The amount we'll accept as "enough" motion in a short time interval becomes our timing threshold. The fan should wait for the value to fluctuate frequently as an indication that people are actually moving in the room. This is essentially the same calculation performed by the sensor itself, but on the scale of seconds, not milliseconds. Since we're using the PIR data to turn on a big, slow fan, checking for several signs of motion over several seconds will ensure the fan doesn't turn on and off constantly.

Once we've seen that period of consistent movement, we need to mark the room as occupied and stop checking as frequently. Again, as in a bathroom, we accept that people will move around when they arrive in a room and sometimes sit for a while. Our code needs to allow the imaginary little people to relax on the couch for a reasonable amount of time without having to wave their tiny arms.

It's essential that we check and tune the three discrete subsections shown in Figure 6-21 in isolation before combining them into a larger program. When debugging a complicated system, all the individual components must perform consistently. Test how much motion it takes to trigger the PIR sensor. Run a Say loop with the temperature sensor and ensure that the fan actually lowers the room temperature. Test all the likely scenarios, and as many strange ones as you can imagine.

Only when the fan, PIR sensor, and temperature sensor behave well in all those tests can we check all three together. When a program evaluates and compares many different inputs, the order and timing of how we check those inputs matters!

As the complexity of mBlock programs increases, it's important to remember that the goal is functionality, not correctness. Rather than worrying about whether a particular solution represents the "right way," keep focused on if it accomplishes your stated goals. Robots and programs are tested, not graded.

Testing throughout the build process not only compartmentalizes large tasks, but the tests force us to consider the question,

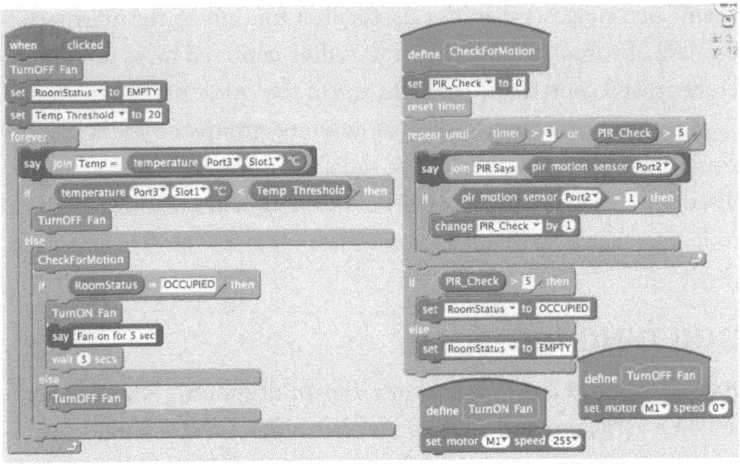

FIGURE 6-21: Once they are all combined into a single stack of blocks, it's trickier to find and correct errors.

"What should happen here?" on increasingly granular levels. That mindset helps even when you're reading unfamiliar programs for the first time—something Makerspace and computer science teachers do on a daily basis. Working through a small part first, and thinking through what should happen at each step, can be very helpful.

We pulled CheckForMotion out as a custom block to isolate the choices made in that process. CheckForMotion looks at the data from the PIR sensor for three seconds or until it sees five motion readings, whichever comes first. The PR_Check value starts at 0 each time CheckForMotion runs, and then increases each time the PIR sensor reports a 1. At the end of CheckForMotion, the RoomStats variable is reset to either Empty or Occupied.

Reading the details of a program should suggest ways for you to extend its functionality or even ways to reshape it around different assumptions. Maybe it's time to add a thermostat sized for tiny LEGO hands rather than use a fixed temperature threshold. Instead of bouncing between off and full power, maybe the distance between the current and ideal temperatures should determine the fan's intensity. Gary Stager names this improv-like questioning process "...and then?" and suggests that it's a useful filter for finding the unexpected corners of complicated tasks. Static adult-centered tasks rarely generate great "...and then?" responses. On the other hand, one or two "...and then?" questions asked of different groups of young people could push this basic project in radically different and fascinating directions. Great projects can generate enough interesting responses to "...and then?" to fill up a whiteboard.

BUILDING BIG

As we've seen in previous chapters, the mBot system allows us to create on a very large scale. Using serial or Bluetooth for wireless communication means that the mCore board or mBot robot can operate far away from the computer or tablet. Uploading the programs to the board, along with the use of a large LiPo battery, allows our creations to operate independently for hours or days at a time. Cheap custom cables make it possible for sensors and motors to spread out along the ceiling or windows of even the largest rooms. In many ways, we've already been "building big" with the mBot.

Now it's time to cross the final threshold of "big" and work with large power loads.

In the section "Connecting Motors with Two Wires (Two-Pole Motors)," we demonstrated how to connect any simple DC motor to the mCore's motor output pins, allowing us to control fans, pumps, and more. All of those devices were pretty small, and easily powered by the mCore's 5V low-amperage output. This works great for prototypes, where exploring and refining the idea is more important than doing real work. But if we want to soak a real kitchen, we'll need to control devices that require far more power.

Controlling large loads with small voltage signals is a central pillar of the Arduino universe. All microcontrollers operate on either 3.3V or 5V electricity. Powering a device through a microcontroller requires all the power used by that device to flow through the same circuit. Controlling larger voltages will require an extra component to switch on the bigger power stream in response to signals from the microcontroller.

There's a whole world of options that will allow you to control exactly the device you want with a given input, and there's no better book to start with than Charles Platt's now classic *Make: Electronics: Learning Through Discovery, Second Edition* (Maker Media, 2015). His hands-on walk-through of physical and solid-state relays and transistors is an essential experience for all Makers.

In our classrooms, we choose parts that are often overkill for the specific applications. When selecting relays or transistors, we stock a few parts that can control a wide range of voltages, rather than finding the switching circuit that's just big enough to handle a specific task. While there's surely a more efficient and possibly cheaper solution for the systems shown in this chapter, that's not our primary concern. Because we allow young people creative autonomy, I'd rather have a dependable and flexible tool ready at hand than go digging through a drawer of parts.

Fire Management System—Large

If you harbored safety concerns when kids were using a candle in a cardboard kitchen, the idea of taking that project to life-size may be downright terrifying. There are plenty of great, safe, empowering ways to teach young people how to build and control fire, but all those lessons are better suited to a camping trip rather than a Makerspace.

Instead of scaling up the fire, we'll focus on scaling up the response. We're going to leave the small pump behind and build a garden hose–powered sprinkler.

Water is heavy, and moving a lot of water requires a corresponding amount of power. That's why most real-world sprinkler systems don't use electric motors to push the water. Fire safety systems rely on water pressure and use valves that degrade and open in extreme heat. Gardening valves, which start and stop the flow of water with solenoid-driven plugs, provide a better model for our project.

In this project, we'll build a water control system that opens a solenoid valve using 12V DC power. Since that much voltage would fry our poor little mCore, we'll need to use an external 12V power supply, and a physical relay. Signals from the mCore will tell the relay to close or open the larger electric circuit that, in turn, controls the valve.

For big DC projects that don't require millisecond-speed switching, I reach for SparkFun's Beefcake Relay board, shown in Figure 6-22.

This board can switch up to a 3A load at 28V DC. The actual relay could handle up to 20A, but the screw terminals and traces on the board aren't rated for huge current loads. The Beefcake can also switch 220V AC loads, meaning that it could handle wall current from most countries, but we don't use the Beefcake for that. Even with the precautions we take in our Makerspaces, having wall current move through a board with exposed terminals and traces gives me the willies. When we need to control something that plugs into a wall outlet, we reach for the PowerSwitch Tail, which we'll use in the room-scale fan project later in this chapter.

Figure 6-23 shows the relay used in the Beefcake relay board with the black plastic housing removed. Don't do this! The plastic cowling prevents fingers from coming into contact with high voltage. Relays use a large copper coil as an electromagnet. When a low-voltage current flows through the electromagnet, the electromagnet pulls

FIGURE 6-22: SparkFun's Beefcake Relay board

a switch closed, which will physically complete a high-voltage circuit. When the current stops flowing through the electromagnet, the switch is released, and the high-voltage circuit is broken.

We'll control this coil with a signal from the mCore to the side labeled Low Voltage. We'll be using this relay to turn something on, which means we'll use the normally open (NO) pins on the high-voltage side. Normally closed (NC) operation means that the circuit is closed by default and the device is powered, except when the microcontroller sends a signal. In our sprinkler setup, an NC relay would be a particularly wet and unpleasant way to fight household fires.

I was first introduced to the Maker utility of sprinkler valves by Joey Hudy's classic marshmallow cannon project (*https://makezine.com/projects/extreme-marshmallow-cannon/*). Garden use valves consist of a connector between two pipes with a solenoid-controlled gate in between. Most are NC, meaning that the plunger blocks the flow between the two sides and requires current to open. Automatic garden-sprinkler solenoids are designed to use 24V alternating current (AC), but can operate for short periods with 12–18V DC

FIGURE 6-23: The coil is at the heart of the Beefcake relay. Image courtesy of SparkFun (https://learn.sparkfun.com/tutorials/beefcake-relay-control-hookup-guide). Image is CC BY-SA (https://creativecommons.org/licenses/by-sa/4.0/).

power. The trade-off when using DC power is that it draws a higher current and generates extra heat while the solenoid is powered and the valve is open. This can cause problems when the whole package is buried beneath the lawn and stands open for half an hour at a time, but won't pose a problem when open only briefly, as it is in this project.

While sprinkler parts from Home Depot or salvaged from the shed will work fine for this project, we will use a 12V DC solenoid valve from SparkFun. Since it's built for light-duty applications like this, it's a bit cheaper than parts from the garden department. It also has a reasonably square bottom and sits nicely on a bench. Gardening valves are designed to be buried in dirt, not sit flat on a work surface.

We'll use the relay to control power to the solenoid valve. (See Figure 6-24.) Only when the relay is engaged and the circuit is closed will power flow from the supply into the valve, which will, in turn, open the valve to allow water to flow.

Figure 6-25 is what would pass for a planning sketch of the sprinkler control circuit in our Makerspaces. When you're working with new tools and materials, abstracted circuit drawings can

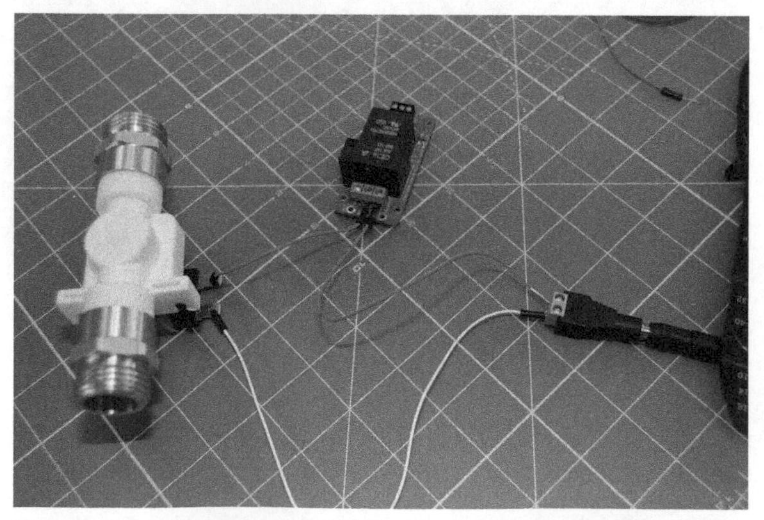

FIGURE 6-24: The power circuit between the 12V source, the sprinkler valve, and the Beefcake relay board

pose real challenges. Although the power plug isn't connected to anything and the tiny wires aren't soldered to the valve, this model is more similar to building the actual circuit than a drawing would be. Modeling with real parts also makes it easier for a teacher to quickly offer feedback.

However, this version is just a model. The "production" build needs some larger, longer wires. Thicker wires are better for higher current loads, and we need plenty of distance between our replay board, the flowing sprinkler head, and the microcontroller.

To keep this type of build flexible, it's a good idea to build small, modular cable connections, as seen in Figure 6-25, discussed at the beginning of this chapter. In this project, we used barrel plugs backed by screw terminals. These plugs are familiar to kid hands and stand up to a lot of stress.

Although longer wires help, the relay board needs a bit more protection. In group or classroom environments, we often package the relay boards in disposable plastic food containers, with small openings to allow access to the low- and high-voltage connections (see Figure 6-26).

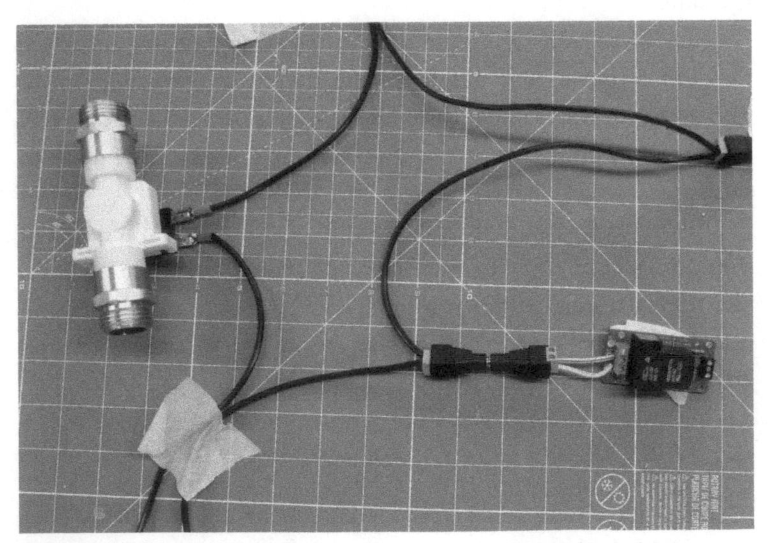

FIGURE 6-25: Similar to Gary's midpoint connectors, but with heavy-gauge wires for high-current applications like a solenoid

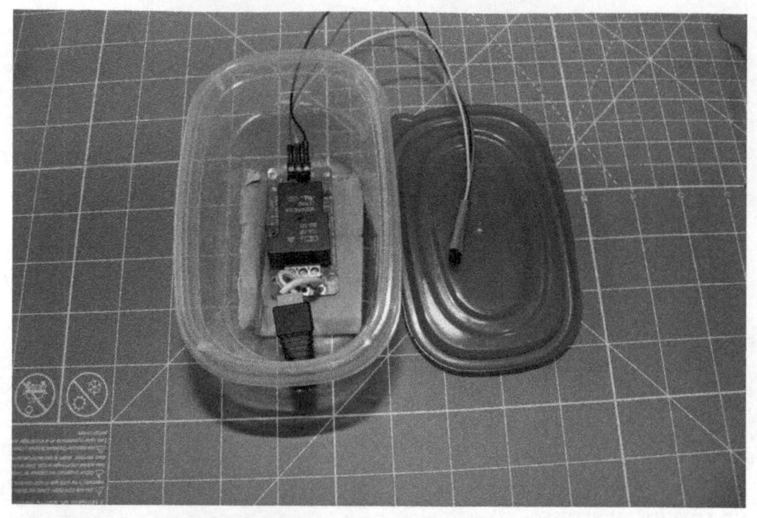

FIGURE 6-26: The power input and the control wires emerge through small cuts in the plastic tub.

This isn't waterproof, but it is splash-resistant. It also presents clear physical instructions to novice users—to control this relay, the only parts that are required are the wires sticking out of the side.

Now we're ready to revisit our fire sprinkler code and modify it for the new parts. At dollhouse scale, the mCore provided power directly to the water pump through the motor ports. Since the relay board requires minimal current, we'll use one of the RJ25 ports to send the control signal instead.

All of the blocks in the mBot section of mBlock are built around a specific sensor or actuator. Accessing more generic commands, like setting a single output to High or Low, requires the Arduino extension. Make sure the Arduino option is selected in the Extension menu, as shown in Figure 6-27.

Then find the Set Digital Pin Output As block in the Arduino section of the Robots palette. (See Figure 6-28.)

FIGURE 6-27: Small check marks in the Extensions menu indicate which blocks appear in the Robots palette.

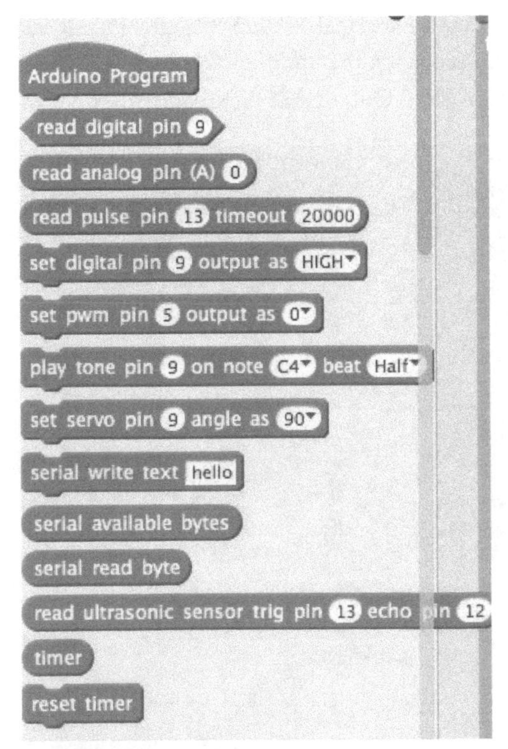

FIGURE 6-28: Programs in mCore can use blocks from the mBot and Arduino extensions simultaneously.

Every RJ25 port on the mCore has wires for two Arduino pins. When you're using the RJ25 breakout board, those two pins are separated into the two 3-wire connection points. To determine which pin number to use in an mBlock program, plug the RJ25 breakout board into a port on the mCore, and then look at the labels behind the mCore connector. (See Figure 6-29.)

In this code we used pin 9, which comes from port 2 of the mCore, and slot 1 on the RJ25 breakout board. Since any of the mCore pins would work for this example, why don't we just use the default values? This means that the signal wire that's connected to the Beefcake relay board traces back to pin 9 on the Arduino, running through port 2 on the mCore to slot 1 on the RJ25 board.

There are a few other alterations to make. The Set Digital Pin blocks call replaces the Set Motor blocks in the SprinklerON and SprinklerOFF procedures. Set Digital Pin 9 Output As High replaces Set M1 to

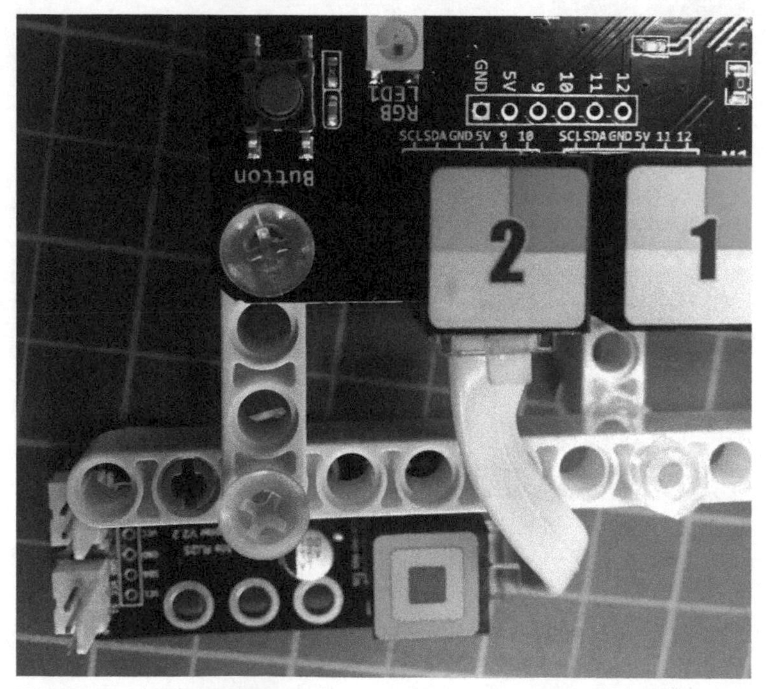

FIGURE 6-29: Pin names are listed behind each mCore port. When the RJ25 board is connected to port 2, slot 1 connects to pin 9 and slot 2 connects to pin 10.

255 in SprinklerON, and Set Digital Output as Low replaces Set M1 to 0 in SprinklerOff. We also removed the Eat sound effect, because it's just hard to hear the computer speaker when you're testing outside. (You can see these changes in the following image.)

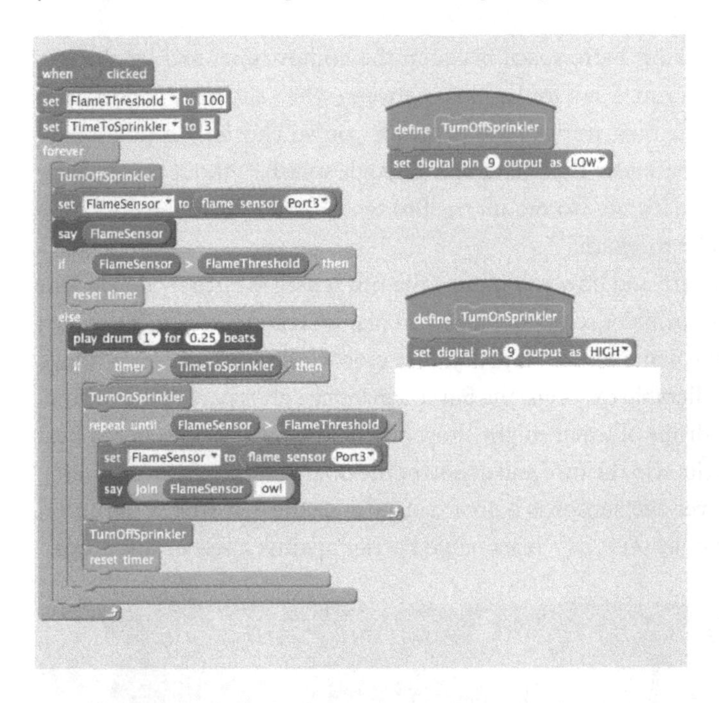

Now we can use this simple code to check the program and wiring of our circuit. The SparkFun Beefcake relay board features a small LED that indicates when the high-voltage side of the relay is engaged. Physical relays also make a distinct and pleasant clicking sound as they open and close. These small visual and audio cues are useful when testing your relay setup. Test your wiring and relay setup before attaching the hoses.

How you position the sprinkler, hoses, and wires depends more on the space getting soaked than the hardware that's used. Pay attention to the threading on the valve and any hoses or connectors. Even when the connectors are the same size, pipe threading requires an adapter for normal garden hoses. Appropriate parts are available at most hardware stores.

Figure 6-30 shows the full setup with the plumbing equipment and the electronics connected, placed artificially close in order to fit in the image. Do not put your electronics this close to the hose and valve. When we ran preliminary tests, the exposed mCore sat far away from the water and under a towel.

Put a bit more space between the components and let the test program run. Look for lag time between when the relay triggers and the water flow stops or starts. Once you've chosen a time interval for the sprinkler that generates enough splash without overloading the relay, it's time to mount the fire sensor and update the dollhouse sprinkler program.

Where and how you mount the fire sensor is entirely dependent on the sprinkler setup and how you plan to test. However, it's crucial that the flame sensor stays dry! Not only would a wet sensor produce unpredictable readings, the flame sensor circuit is part of the mCore. Stray drops of water might short circuit the mCore and, in the best scenario, trip the fuse and depower the board. Adding a layer of cling film over the sensor is a great safety measure—it isn't waterproof, but it does serve as a reasonable barrier against a few errant drops.

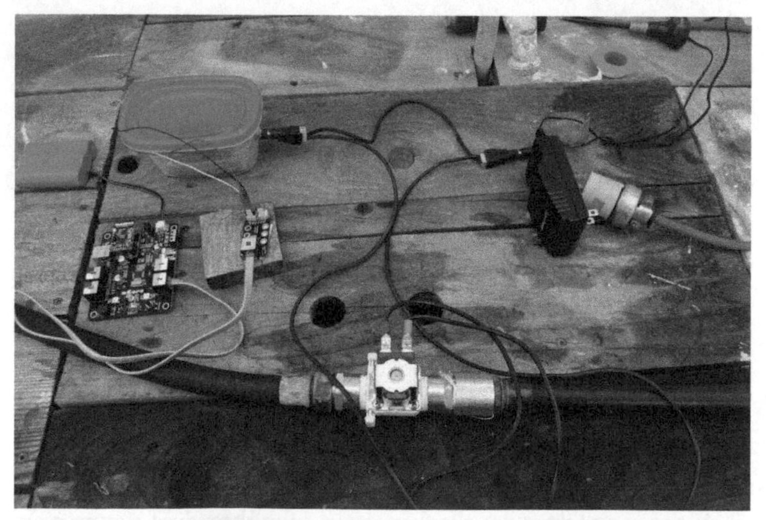

FIGURE 6-30: Here's the mCore and battery, connected to an RJ25 board, which is connected to the Beefcake relay board, which is housed inside a waterproof tub.

Once the flame sensor is mounted, make sure to test the reading. A frustrated kid jumping around with a lit candle trying to trigger the sprinkler increases the risk of this project significantly. Thanks to the wide angle of the sensor, the example setup will detect a flame in a large arc anywhere between 3′ and 8′ off the ground.

Now, gather your volunteers, and get ready to test! Figures 6-31 and 6-32 shows before and after.

FIGURE 6-31: Wait for it...

FIGURE 6-32: It works!

Although this sprinkler setup emerged out of the dollhouse fire alarm, it doesn't have to end there. You now have the ability to program arbitrary reasons to soak people! Maybe that traffic light classroom volume monitor was too passive. Soak the loud ones! Maybe balloon jousting should end by drenching the losing team. The possibilities are endless . . . and soggy.

Fan for Crowded Room—Large

After that mess of relays and hoses, it seems like scaling up the PIR temperature fan project should be much easier. The PIR sensor has a huge field of view and can easily cover most of a room. Discretely hiding a thermometer is easier at human scale than in a cardboard box. It seems like scaling up this project is just a matter of adding longer cables—until we get to the fan.

Even 120 mm computer fans don't move enough air to affect the average human-scale room. However, box, desk, and oscillating room fans are powered by AC motors, rather than DC. These motors plug directly into local wall current (110V in the United States) to spin big blades and move a bunch of air.

Wall current is super dangerous—deadly, even! As a rule, our Makerspace does *not* work with *mains electricity*, aka, what comes out of the wall socket. Not only does it have the potential to fry people, it would obliterate all of the robots or motors we've seen thus far. Wall current is not your friend!

Room-sized fans need wall current. Although big relays, like the SparkFun Beefcake used in the sprinkler project, can switch 110V AC, we don't use them with kids. Exposed traces and screw terminals present only a mild risk of accidental shock, but that's more than we're willing to accept when we have a classroom of adolescents.

Instead, we turn to the PowerSwitch Tail, shown in Figure 6-33, which encases a high-current relay inside a traditional power brick.

The input wires on the PowerSwitch are fully isolated from the relay and the wall voltage circuit. Although I can't say that this eliminates my nervousness at having young kids working with wall current, it's enough to get the project moving. For budget-conscious electrical experts, there are much cheaper ways to use the mCore to

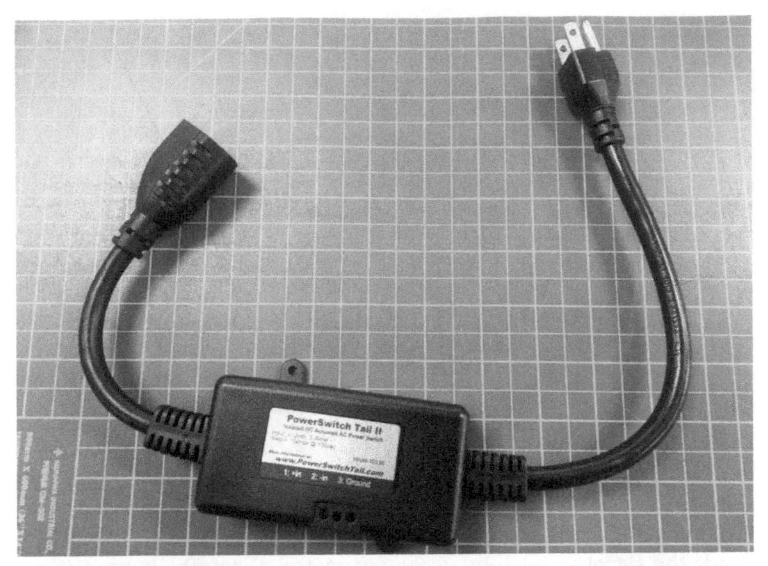

FIGURE 6-33: This PowerSwitch Tail is designed for use with US 110V wall current. There's another version with appropriate plugs for countries using 220V standards.

control large fans, toasters, or hair dryers, but I sleep much better spending the extra cash on these.

The PowerSwitch Tail connects to the mCore like other relays, using the RJ25 expansion board. Since the low-voltage side of the PowerSwitch Tail is *opto-isolated*, meaning there is no physical connection between the high-power and low-power circuits. Instead, the bridge between the two circuits is a tiny LED and light sensor, not unlike the onboard sensor on the mCore. Because of this setup, you only need to connect the signal and ground wires from the RJ25 board, not the 5V wire. Connect the signal wire to the + Input pin, connect the ground wire to the – Input pin, and leave the ground pin on the PowerSwitch Tail empty, as shown in Figure 6-34.

As with the Beefcake, we'll need to use the Digital Pin block from the Arduino extension to switch one particular pin to High or Low. A signal LED on the PowerSwitch Tail shows when the relay is engaged. Instead of having to hack apart a cable, the PowerSwitch Tail sits neatly inline between the room fan and the wall outlet, and the mCore can now control any household lamp or fan.

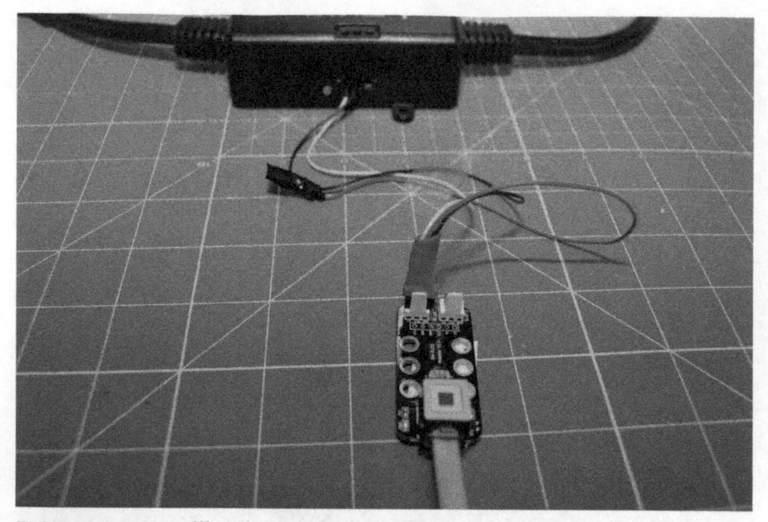

FIGURE 6-34: Unlike the Beefcake relay, the PowerSwitch Tail only needs the signal and ground wires connected. Cover the loose 5V wire from the RJ25 board.

With the PowerSwitch installed, the biggest challenge is replicating the physical setup of the dollhouse room at full scale. This is absolutely the time for super-long extension cables. Just like in the scale model, it's important to test as you go to ensure that the airflow lowers the temperature on the thermometer. Also, the timers for the PIR sensor and the fan will need to be significantly longer. The real world is big, and it takes time for stuff to move around!

These projects represent one way you can extend mBot's capabilities far beyond what arrives in the retail kit—but it's not the only one. The mBot arrived into our elementary and middle school classrooms as an accessible, low-floor, programming and robotics platform. What's kept them in use throughout middle and into high school is that the full power of the Arduino platform is just under the surface. Almost any Arduino project found in an issue of *Make:* will work with an mCore at the heart.

Index

C

D

E

F

G

H

I

J

N

O

P

R

S

U

V

W

X

Y

CPSIA information can be obtained
at www.ICGtesting.com
Printed in the USA
BVHW01s2105261117
501034BV00002B/2/P